D.J. CASLEY AND D.A.LURY

DATA COLLECTION IN DEVELOPING COUNTRIES

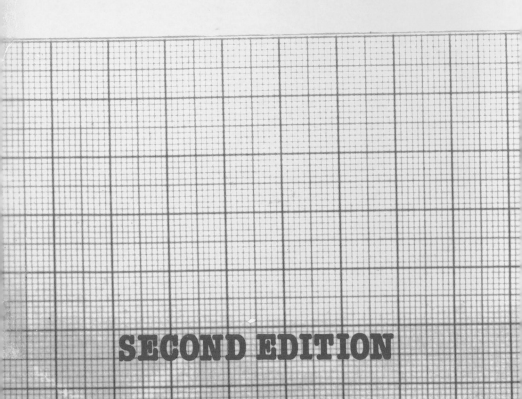

SECOND EDITION

DATA COLLECTION
IN
DEVELOPING COUNTRIES

Second Edition

Data Collection in Developing Countries

Second Edition

D. J. CASLEY

AND

D. A. LURY

CLARENDON PRESS · OXFORD

1987

Oxford University Press, Walton Street, Oxford OX2 6DP

Oxford New York Toronto
Delhi Bombay Calcutta Madras Karachi
Petaling Jaya Singapore Hong Kong Tokyo
Nairobi Dar es Salaam Cape Town
Melbourne Aukland
and associated companies in
Beirut Berlin Ibadan Nicosia

Oxford is a trade mark of Oxford University Press

Published in the United States
by Oxford University Press, New York

First edition first published 1981, first issued in paperback 1982
This edition published 1987

British Library Cataloguing in Publication Data
Casley, D. J.
Data collection in developing countries—2nd ed.
1. Sampling (Statistics)   2. Social surveys—Developing countries.
I. Title   II. Lury, D. A.
300'.723 HA31.2
ISBN 0-19-877283-1
ISBN 0-19-877282-3-pbk

Library of Congress Cataloging in Publication Data
Casley, D. J.
Data collection in developing countries.
Includes index.
1. Sampling (Statistics)   2. Developing countries—
Statistical services.   I. Lury, D. A.   II. Title.
HA31.2.C37 1987   001.4'222'091724   87-12574
ISBN 0-19-877283-1
ISBN 0-19-877282-3-(pbk)

Set by H Charlesworth and Co Ltd, Huddersfield, England
Printed in Great Britain
at the University Printing House, Oxford
by David Stanford
Printer to the University

To Jill and Toni

Preface to the First Edition

Our purpose in writing this book is set out in the Introduction. We decided to write it in Nairobi in 1978 when we renewed a working relationship that had started in Uganda twenty years earlier. Looking back, and reviewing the current situation, we thought that a book concentrating on practical problems of data collection in developing countries would be helpful. We know, of course, that many of the problems are problems without solutions; but continuous effort directed by common sense and experience can do much to reduce difficulties. We have set out what we have learned, in the hope that this will help others avoid our mistakes.

The quotations at the beginnings of the chapters are taken from the cases of Sherlock Holmes, reported by A. Conan Doyle.

Chapters have been read and commented on by our colleagues, A. Bebbington, J. Dobby, M. F. Fuller, M. A. Katouzian, E. Oxborrow, S. C. Pearce, A. Rutherford, and P. Stirling of the University of Kent; and T. Marchant, R. D. Narain and S. Narula of the FAO. We are grateful for their assistance; but they are not, of course, responsible for our errors.

One of us (D. Casley) wishes to thank the FAO of the United Nations for permission to publish this book whilst a staff member, and to acknowledge the use of notes of lectures given on behalf of the FAO. We emphasize, however, that the content of the book is based upon our opinions and experiences and does not necessarily reflect the views of the FAO.

We are grateful to Miss A. Akhurst, Mrs. M. Averdung, Mrs P. Burton, and Mrs M. Weston, who have borne the main burden of typing and retyping our drafts.

Finally, we express our appreciation to those with whom we have worked; in particular the enumerators and field supervisors who collected the data and the respondents who put up with our inquiries and intrusions. We make plain in the book our belief that, to be effective, a surveyor has to work very closely with the field staff. This necessity has given us much pleasure and many friendships.

D. J. CASLEY
D. A. LURY

Canterbury, 1979

Preface to the Second Edition

My collaborator Denis Lury died in 1981 so, perforce, the revisions undertaken for this edition have been made without the benefit of his wisdom and experience. I owe much to his guidance, particularly when I undertook my first surveys in Uganda some thirty years ago, so what good the reader may find in this edition still owes much to him.

I am grateful to my colleagues Miss Diana Crowley, for her assistance in preparing the revisions, and Mr Vinh Le-Si for his review of, and comments on, Chapter 9. Also, I thank my wife, Jill, for typing the revisions to the text during what was ostensibly her holiday.

Finally, I express my appreciation to those who have commented on the first edition: it is pleasing that some have found it helpful in planning and undertaking their own surveys.

Mullion, Cornwall, 1986   D. J. CASLEY

Contents

Abbreviations

Bull. Intl. Stat. Inst.	*Bulletin of the International Statistical Institute*
IBRD	International Bank for Reconstruction and Development
Int. Statist. Rev.	*International Statistical Review*
IUSSP	International Union for the Scientific Study of Population
JASA	*Journal of the American Statistical Association*
JRSS (A)	*Journal of the Royal Statistical Society*, Series A
Phil. Trans. Roy. Soc.	*Philosophical Transactions of the Royal Society*
Popn. Studies	*Population Studies*
Proc. Roy. Soc. Lond.	*Proceedings of the Royal Society, London*
WFS	World Fertility Survey

Introduction

Still, elementary as it was, there were points of interest and
novelty about it which may excuse my placing it upon record.

The Adventure of the Blanched Soldier

Data collection involves a range of activities, from the individual in a library extracting information from volumes of national and international statistics to a team of thousands carrying out a national census. In this book we consider most of that range, covering techniques of inquiry from the case study to the census, but give major emphasis to data collection by sample survey.

Although the basic ideas of sampling are very old, the development of sampling theory is comparatively new, and is one of the major intellectual achievements of this century. It provides a logical conceptual framework by which estimates of the characteristics of a population can be inferred from the results of an examination of only a sample of that population. There are a number of excellent texts dealing with the theory; we do not wish to duplicate them. We discuss the practical aspects of carrying out a sample inquiry, which are sometimes over-looked or taken for granted.

In recent years there has been growing emphasis on survey techniques that are not based on sampling theory. One school speaks of rapid appraisal methods and argues against formal sampling. In large part, this is due to a common misunderstanding that equates a formal sample inquiry with a complex questionnaire applied to very large samples. We attempt to correct this inappropriate equation, but it is disturbing that it seems to be more widely held today than when this book was first published.

The special difficulties of conducting surveys in developing countries derive from their socio-economic structure. These countries are in a period of rapid transition—demographically, economically, and culturally. They normally have high, but changing, birth and death rates. There is great mobility, particularly by rural–urban migration. Agriculture, still the main occupation of most of the population, is one of the major subjects investigated by sample surveys, but it presents particular problems. Climatic and soil conditions within one country may vary to such a degree that agriculture is practised in arid or semi-arid lands at one end of the country while planting and harvesting are continuous throughout the year at the other. Some areas may be inhabited by nomadic pastoralists, others by small farmers, with modern commercial farming interspersed in clustered pockets. Thus there will often be wide regional disparities, particularly since the differences in physical conditions will usually be accompanied by cultural differences. Many areas may be inaccessible for parts of the year that coincide with important periods of the agricultural cycle.

The mailed questionnaire or the diary of events will be practicable in only a few inquiries, such as those involving firms or an educated minority; they cannot be used in general household inquiries into income, expenditure, agricultural practices, and similar topics. An interview based on recall of facts presents major difficulties in most cases. Areas of land may not be known, and in any case may be of little direct use owing to mixed cropping. The concept of a household may well vary from one region to another; in many it will be a nebulous concept, difficult to define. Direct objective observations will often be required, but are costly in terms of survey resources and sample size limitations. Intimate local knowledge is essential.

Clearly, no general solutions can be offered. Our aim is to provide a description of the problems that identifies their common features, and a discussion of the techniques that have been developed to deal with them. Many of these techniques are little more than codified common sense, but it is surprising how often simple aspects are overlooked, sometimes through inexperience, often through a choice of sophisticated techniques that are inappropriate in the context. There is often a failure to appreciate logistic and staffing difficulties; and there is a tendency to follow a design misleadingly termed 'optimal' because it satisfies certain technical criteria, but which is in fact sub-optimal because of the circumstances in which the survey has to be carried out. Although we shall not deal directly with sampling theory, we cannot and do not ignore it. Indeed, one of our major aims is to enable investigators to develop an appropriate marriage of theory and practice.

Our basic message can be summed up easily: it is, 'keep it simple'. One application of this rule is to content. The minimum amount of information required to meet policy needs should be established. The discussions by which this streamlining is achieved are not required just to minimize the work the surveyor has to do; they usually play an essential role in clarifying the issues that the survey is expected to illuminate. Particularly when the survey is a commissioned one, the 'cross-examination' of the commissioner is not just for the benefit of the surveyor: it is also a valuable part of the education of the commissioner, although he* may find it painful at the time.

The second aspect of simplicity is a technical one. The sampling errors

---

*General note: English does not contain a convenient unisex pronoun: we shall use 'he', but this should not be taken to mean that we think roles are sexually constrained. We hope female readers do not feel we are discriminating against them: we used 'he or she' in our early drafts, but it is a very tedious business both for writing and reading.

of any rational design involving at least a moderate sample size are likely to be substantially smaller than the non-sampling errors. Complications of design may create problems, resulting in larger non-sampling errors, which more than offset the theoretical benefits conferred. The same emphasis on the advantages of simplicity also applies to our recommendations regarding the organization and methodology of the survey operations, including the analysis and presentation of results.

We hope that this book will be useful, first of all to those statisticians and research workers in developing countries who, whilst abreast of the theory and aware of some of the difficulties that may occur, have not yet been exposed to the whole range of problems from the initiation to the conclusion of a survey. The book should be of use also to officials, both 'locals' and 'outsiders' concerned with development aid, and to other planners and consultants who use the data collected in developing countries and therefore need to appreciate their limitations. Moreover, such users often have to turn their hand to conducting a survey to fill a gap in the data available or to meet a deadline. Finally, we direct this book at development project staff and research workers in developing countries, who may have only a limited statistical background, but who often need to carry out data collection for use by project management and for research into socio-economic aspects of development. We hope this book will help them to appreciate and overcome some of the difficulties they are likely to encounter, and that it may assist in promoting fruitful discussions between them and any statistician they consult.

We are in no doubt that there are problems that we have not experienced and solutions of which we are unaware. We shall be well satisfied if we have at least mapped adequately some tracks through the morass of difficulties the investigator will encounter.

1
Collecting Data: A Historical Overview

'These relics have a history then?'
'So much so they are history.'
'I should be glad' I said, 'if you would give me an account of it.'

The Musgrave Ritual

1.1 INTRODUCTION

Data can be collected from a defined population by recording the appropriate information about every member of that population. This is a *census*. Alternatively, data can be collected for only some of the members of the population. To describe this process we shall refer to the selection of a *sample* and the operation of a *sample survey*, or merely to a *survey*. The phrase *sample census* is sometimes used to refer to regular (often decennial) major efforts to collect data regarding the composition and structure of a population which do not cover every member of the population. We shall avoid this term as many find it confusing. Data collection requires both censuses and sample surveys. In the early years of building a national data collection capability, the emphasis was on the census approach, which restricted the range of data that could be included to population counts and the enumeration of well-defined statistical populations, limited in number and easily accessible—for example, censuses of manufacturing industries or censuses of licensed traders. Usually, only government organizations have the power and authority to carry out censuses. With this emphasis large gaps inevitably exist in the information base. In the agriculture sector, census data for large farms may be available on an annual basis, with no data at all for the much larger number of small farms, which in aggregate contribute more to the economy. In some cases data are compiled and presented as if based on complete enumeration when, in fact, the recording process omits much of the population. Data on births and deaths registered reflect only a small fraction of such events, the existence of rules requiring such registration notwithstanding.

To fill these gaps and repair the deficiencies in such incomplete coverages recourse to sampling techniques is required. For non-official agencies requiring particular data on the population at large there is no alternative to sample surveys. And for individual researchers the case study, involving a detailed study of a few members of the population, may be either the most appropriate method of study, or all that can be done:

formal inferences from the few 'cases' to the population as a whole will not then normally be involved. Alternatively, an individual research worker may use the data collected by others, supplemented by personal observation. Such informal observation in the rural sector has been codified to some extent under the rubric of rapid rural appraisal.

Censuses and surveys can suffer from the same types of error in enumeration; these may be grouped into errors of coverage and errors of content. Members of the population, or the selected sample, may not respond, giving rise to errors of coverage or non-response. Content errors may be caused by falsification or misunderstanding on the part of the respondent. Similar errors may be made in recording by the enumerator owing to his misunderstanding, incompetence, or dishonesty. Further errors may arise in the analysis and presentation of the results. In addition to these non-sampling errors, sample surveys suffer from one source of error that does not occur in a census, namely that resulting from sampling variability. The sample actually selected gives a result different from that which would arise from another sample of the same size chosen in the same way from the same population. This additional error may be offset by response or enumerator errors being larger in a census than in a sample survey. These larger errors can arise in the census because the bigger scale of operation results in a lower standard of enumerator, less efficient training, and greater problems in supervising the data collection process. Justification for a census must rely on the vital need to know details of the population at a very low area level. Population censuses, even if little more than head counts, may be justified on this basis as the population within small administrative units may be needed for several important purposes. We consider there are very few other inquiries that require a complete enumeration. We do not accept as a justification the need for accuracy; in virtually all instances we have encountered, response and observational errors increase rapidly with an increase in the number of respondents, and are of a higher order of magnitude than the sampling error. To aim for complete enumeration, or even very large samples, will often bring about less, not more, accurate results.

With censuses limited in their feasibility, it was the application of the sampling method that provided the keystone for the development of data collection in the developing countries. It is worth reviewing how recently sampling theory was established.

1.2 THE ADVENT OF SAMPLING

All people experience sampling in their lives. Our knowledge of the past is based on samples; samples of the conversations of our elders, samples of the books, music, works of art, and systems of thought that have survived, almost certainly in a biased fashion, from the past. Historians examine a sample of a sample in order to find or impose a pattern on the past. Our knowledge of the present is similarly based; our views of people we know come from our knowledge of a sample of their actions and attitudes, and our opinions of countries or institutions are based on our knowledge of a sample of the actions of a sample of their members. These statements are truisms, but suitably emphasize at the start of this book that everyone, every day, is acting as a sampling statistician in that decisions are made and actions taken on the basis of knowledge obtained from a sample.

The use of sampling techniques in a conscious fashion—the deliberate selection of a few units from which it is intended to draw conclusions about the whole—goes back a long time. To judge the quality of bags of corn by sample handfuls, the quality of a roll of cloth by the inspection of sample lengths were, and still are, commonplace actions of the market-place. Customs officers in former days fired volleys of musket shots into wagon loads of hay to see if they concealed smuggled brandy. However, sampling in this simple manner frequently led to wrong conclusions, so when national statistics offices and statistical branches of other institutions began to develop in the nineteenth century, the sampling method was regarded as an unreliable method of data collection. The exploitation of the sampling process waited upon the development of an appropriate theory, and a recognition of the way in which it could be applied.

The crucial theorem on which modern sampling theory is based can be traced back to the late seventeenth and early eighteenth centuries; but its full development, with the analysis of the Normal or Gauss–Laplace distribution and its relation to the 'law of errors', occurred around 1800. Suppose we imagine a large population of items of varying size. There will be a mean or average size obtained by adding all the sizes together and dividing by the number of items. If we consider taking successive samples of a stated number from this population we shall find that a few samples consist of the small members of the population and so produce a sample average well below the population mean, whereas other samples consist of the large members and so give rise to a sample average well above the population mean. Most samples, nevertheless, will include a mixture of larger and smaller items and will give sample averages close to the population mean. Provided that the samples are reasonably large, the distribution of their averages can be approximated by the Normal distribution.

A long time passed before this work was used to provide the foundation for a theory of sample surveys. During the nineteenth century the census reigned supreme. In a discussion about the 'representative method', initiated by A. N. Kiaer in 1895 at a meeting of the International Statistical Institute, all the counter-arguments were based 'on the alleged sanctity of the census method'.[1] Bowley, speaking in 1906, said that 'the relation of the frequency of deviations to the law of error was regarded as a statistical curiosity' and that 'mathematical methods of testing the truth of practical deductions have as yet borne singularly little fruit'.[2]

In the 1895 discussion, Kiaer described work he had carried out in Norway. The inquiries included features that are still very much part of current practice, especially the use of stratification and the varying proportions of different strata selected for study. The main difference between his procedures and those recommended today is that the choice of the final sample units does not appear to have been precisely specified and, at least in some cases, this final selection was left to the discretion of the enumerator.

Kiaer monitored his samples by comparing sample averages or proportions for certain characteristics with similar information obtained from a previous census. The question that then arose was how to find objective grounds for deciding how close the sample and census figures for these characteristics had to be before accepting the sample figures for those characteristics for which census results were not available. Bowley circumvented this issue. He considered the distribution of sample averages derived from samples selected by the method of simple random sampling, when the survey is designed so that each unit in the population has the same chance of selection. He argued, 'If quantities are distributed according to almost any curve of frequency satisfying simple and common conditions, the average of successive groups of say 10, 20, 100, ... n, of these conform to a normal curve (the more and more closely as n is increased) whose standard deviation diminishes in inverse ratio to the number in each sample.'[3] Bowley showed that a standard deviation calculated from only the sample information can provide a measure of the precision of the sample average. This was the decisive step and Bowley and his colleagues employed the technique in a number of surveys over the next twenty years.

Its use did not spread rapidly, however. In 1924, the International Statistical Institute appointed a committee to study 'The Application of the Representative Method in Statistics'. (At this time the word 'representative' covered methods involving random or purposive selection. In later usage, 'representative' is often used as synonymous with 'purposive'.) The committee's report did not rule out purposive selection entirely; it was left to Neyman in his paper of 1934, 'On the Two Different Aspects of the Representative Method',[4] to show that inferences could not be drawn in a satisfactory fashion from purposive samples, but only from samples where the selected units were chosen with a known probability. This paper also placed stratified sampling on a sound theoretical basis, and derived the basic relationships from which the optimum allocation of sampling units to strata could be obtained.

Parallel with these developments, Fisher was laying the foundations of modern statistical theory. Building on the work of Student—the pseudo-nym of W. S. Gossett—he was showing the way in which samples of the limited size practicable in experimental work in agriculture, medicine, and the natural sciences could be used in making choices between hypotheses.

Sampling was still not generally accepted, however; in the mid-1930s in the USA 'sampling was trusted neither by the public nor by members of Congress; the idea was both innovative and perhaps "far out" that the Census Bureau, which based almost all of its work on complete coverage or censuses, might make use of sampling methods. A common view of that time was that others could take such risks with loose methods, and thereby undermine the basis for confidence in their data, but not the Census Bureau.'[5] Prejudices such as these were finally convincingly refuted by the successes of national statistical activities in India, the United States, and the United Kingdom.

In India, P. C. Mahalanobis, a close colleague and friend of R. A. Fisher, developed a large-scale sampling scheme for the first exploratory survey of the cultivated area and yield of jute in Bengal. In two classic papers[6] he outlined the theory and practice of large-scale surveys. He paid particular attention to cost functions, interpenetrating samples, and to practical problems of enumeration and analysis; he could report in 1943 with justifiable pride:

To judge the success (or otherwise) of the scheme the Jute Census Committee had laid down three tests. The reliability of the sample survey must be such that the margin of error of the final estimate of the area under jute should not exceed 5 per cent; secondly, the results must be available sufficiently early in the jute season and preferably by the first or second week of September; and finally the cost of the sample survey should not be excessive . . . the sample estimate of 1941 agreed within 2.8 per cent of an entirely independent official estimate based on a complete deailed census . . . [it] was submitted to Government on 27 August . . . . The cost of a sample survey was estimated at about £8,500 against an expenditure of about £110,000 for a complete census.[7]

Similar developments were taking place in the United States, arising out of the 1937 Enumerative Check Census on the Census of Unemployment. Sampling was also used in the 1940 Population Census and was the basis of the 1943 Labor Force survey, which helped to shape US national man-power policy in the war, and so helped to gain acceptance for sampling in the administration. Cluster sampling was used in Bureau of Census work, and its variance was dealt with in a paper by Hansen and Hurwitz.[8] The similarities in the problems and the methods developed to deal with them were clear to the statisticians of India and the USA when they met.

The growth of activities in the UK was more diffuse, and the pioneering work of Bowley and his colleagues had already built a store of experience. Wartime needs forced the extension of sampling, and a series of inquiries were made, those in agriculture being influenced by F. Yates and the Fisher school at Rothamstead Experimental Station.

One of the first statistical activities of the United Nations was to set up, in 1947, a Sub-Commission on Statistical Sampling Methods, involving Fisher, Mahalanobis, Yates, and Deming. Yates was asked to produce a manual which was published in 1949.[9] Another valuable text, by Deming, appeared in the following year.[10] Although there have been many developments since, the essential processes for efficient sampling were widely available and authoritatively supported from then on.

1.3 THE DEVELOPMENT OF STATISTICS AND THE STATISTICS OF DEVELOPMENT

It might be thought that the statistics offices of the developing countries that were being created at that time would turn to this conveniently available sampling theory and develop data collection around some form of national sampling plan. After all, sampling techniques were said to be, amongst other things, economical in time, labour, and cost. What better choice for the needs of poor countries? In general, although some individual surveys were carried out, a coherent adoption of sampling did not take place for a variety of reasons.

First, sampling requires frames from which to sample and these were unavailable until censuses were conducted, providing clearly and fully documented lists of administrative locations. Secondly, despite the new textbooks, knowledge and experience of large-scale surveys were limited. Many of the statisticians working in developing countries in the 1950s (and there were often no more than two or three in any one country) had an economic rather than a mathematical background. Such mathematical statisticians as there were tended to be in research institutions and so concentrated on experimental design and the application of analysis of variance techniques. In addition, the infrastructure for carrying out survey programmes in the field was not available; staff in sufficient numbers at the junior levels did not exist.

The only exception to this situation in developing countries was the extension of Mahalanobis's work in India which led, in 1950, to the creation of a National Sample Survey. This was possible because of the existence of cadastral maps on almost a nation-wide scale and the flow of able mathematical statisticians from within India provided particularly favourable circumstances—in fact, in statistical terms, India was highly developed, with its long historical background of pioneering statistical work.

Besides, there were some areas of key importance, particularly in the economic field, in which administrative statistics (statistics collected as a by-product of the administrative process) were available on the basis of complete, or nearly so, enumeration, and were thought to be of reasonable quality. For example, import and export statistics were generally available from documents submitted to and summarized by customs departments. There were good grounds for believing that the limited statistical resources could make a wider and more immediate impact by exploiting sources of this kind. This expectation was also justified in terms of influencing the policy makers. Most of the senior officials concerned still took a critical, albeit uninformed, view of samples. But they were much less sceptical about information derived from censuses or from administrative sources. This was not surprising, as usually they, or people like them, had been involved in the production of such information.

At times the reliance on administrative sources became dangerously misleading and, in view of the potentialities of surveys, almost ludicrous. For information on areas under crops many countries relied on returns submitted by local chiefs, coupled with an estimate of mean plot size, which was sometimes based on a few recent measurements by agricultural officers, but was often subjectively derived and maintained as a constant over the years. We once reported to the head of an agricultural department that the results of a sample survey suggested that the official estimate of the area under an important cash crop was grossly overstated. His suspicion of our methods was bolstered by his subjective experience, and his reply, 'I have walked all over this country for twenty years, I've seen these fields,' provides a good example of the reluctance to rely on sample estimates. Not until the third independent sample survey some three or four years later had confirmed the other two in indicating gross overestimates in the chiefs' returns did the department accept the new estimates. Since the overestimate of area had resulted in an underestimate of yield, it had provided a bad guide for policy in the interim.

Perhaps the most decisive factor at this time was the importance attached to national accounts—above all to the calculation of estimates of the Gross Domestic Product. There was a statistical justification: part of the case for the central role of the national accounts calculation was that it provided the most convenient framework for fitting together existing data and assessing the relative priorities of various statistical activities. The main reason for its importance, however, was that the form in which it had been developed in Britain and the United States during the war was designed to provide the orders of magnitude occurring in the key relationships of the Keynesian economic framework, which was being adapted for national economic planning models in many countries.

The general method of calculation relied on administrative statistics of external trade, public finance, agricultural department estimates of crop and livestock production supplemented by marketing board reports, population data, and specially collected estimates of employment and earnings in a sector usually designated as 'modern' or 'monetary'. The GDP calculation has retained its priority status, and much effort has been devoted to its extension and improvements. Constant price calculations, and functional and economic classifications of government expenditure, are made. Sampling was at last used in order to obtain supporting information about the income and expenditure of urban workers.

Details of the important transactions and interrelationships occurring in the mass of the population working in agriculture or in informal activities continued, and in many countries continue, to be neglected. Although sampling is increasingly used for such surveys as are conducted into this general population, the standard of data collection is often poor, leading to indifferent results and a consequent lack of support for this type of inquiry.

The needs and resources of developing countries require that official statistical policy should now give greater emphasis to investigating the activities of the mass of the population, that is, the activities of the agricultural and the rural and urban information sectors. We do not suggest that the improvement of the calculation of the GDP should cease; indeed, it is likely that the emphasis we suggest will, amongst other benefits, refine the calculation, particularly in those areas for which data have always been deficient in quantity and quality. It seems strange that statistics offices that cannot estimate with reasonable accuracy the number of livestock at one point of time—let alone annual changes in the number —should devote scarce resources to refining classifications of minor items in public expenditure. A change in priorities will necessarily result in some reorganization and the reallocation of resources.

One reason why these small-scale but widely spread activities have not been covered is that their very nature makes it difficult to collect information about them. Circumstances will of course vary from country to country. It is easy to say that the key is to achieve the most appropriate mix of censuses, surveys, and secondary sources of material. We hope that the review of the advantages and disadvantages of different methods of data collection given in this book will provide a useful guide to a coherent total programme.

The correct assessment of priorities is of prime importance to the government statistician, as it is to managers in any field. For some time to come there will be more major topics deserving investigation than there are resources (both trained manpower and financial) to cover them.

The major data collection effort of any government statistics bureau is likely to be concentrated on the big 'set pieces': the census of population, the surveys of business and industry, and the surveys of agriculture. The structure and composition of the population and economy, as revealed by the data from these inquiries, provide the backcloth against which all other data will be assessed and evaluated. There is no need to set down the claims for priority for these inquiries; their essential nature is recognized, although in the case of agricultural surveys recognition has not always resulted in action. Rather, the need is for a word of caution. Important though these investigations are, they should not be allowed to absorb all the available resources. A decennial population census must feature high on any priority list, but if the census leads to a succession of surveys of population change, fertility, mortality, etc., a situation can be reached where too much of the total resources goes into demography, starving other major areas of study.

The study of the agricultural sector must be near the top of the priority list, but as already mentioned, this sector is often the least surveyed. One reason for this is the expense of conducting national agricultural surveys in situations where the postal inquiry or even direct interviews will not suffice. Often objective counts and measurements by a trained field force are required. Although the composition of the agricultural population, the structure of holdings, and methods of land utilization may be suitably measured once in ten years, some of the most important data relating to agriculture must be collected annually if they are to have any value. Population change between census years can be closely estimated, but agricultural production may fluctuate sharply from year to year as the result of climatic and market influences. More resources are needed and the scope and contents of agricultural surveys need very careful formulation if the best use of resources provided is to be made.

Surveys of the business and economic sectors are of major interest to most governments. Fortunately such surveys, although demanding in terms of skilled statistical inputs, may often be conducted for modest financial outlays since postal surveys, with their savings on enumerator manpower, may be mounted.

If the resource needs of the big 'set piece' surveys are kept under careful control, resources will be released for conducting surveys of other areas, or more in-depth surveys of certain aspects of the general demographic, agricultural, and industrial sectors. Deciding priorities for these *ad hoc* or occasional surveys is a delicate matter. In determining the programme and timetable for the big standard surveys, the statistician may take the leading role, partly because everybody agrees they are necessary; argument will be about what goes into them. The statistician must be guided in the allocation of remaining resources by the need of the planners and decision makers in the various government ministries.

1.4 RECENT HISTORY AND FUTURE REQUIREMENTS

In the last two decades considerable attention has been given by governments and international agencies to improving the quality of demographic data. Population censuses of reasonable quality are the norm, and the surveys sponsored by the World Fertility Survey programme resulted in a considerable corpus of knowledge on the dynamics of population growth.[11] Many countries have conducted *ad hoc* urban, regional, or nation-wide household surveys to collect data on the variables affecting household affairs—income, expenditure, consumption, labour, health, etc. Under the auspices of a UN-sponsored Household Survey Capability Programme, attempts have been made in the 1980s to establish a permanent capability for such surveys in participating countries, but the problems have been considerable and achievements limited. The FAO, which has for many years promoted a World Census of Agriculture[12] on a decennial basis, more recently has emphasized the need for annual sample surveys of farm inputs and outputs; but, again, the establishment of such a capability has proved difficult. Most recently of all, many countries, encouraged by WHO and UNICEF, have instituted anthropometric surveys of young children to determine their nutritional status.[13] This has been one of the more promising recent developments in data collection.

Meanwhile, there has been an explosion in data collection activity by national agencies that are not directly responsible for official statistics. Most of this has been motivated by the need to monitor and evaluate specific development projects—particularly agriculture and rural development projects.[14] Such efforts have often been well financed, but lacking in staff of proven survey experience; consequently the success rate of these has been moderate at best, with several expensive fiascos among them. Whether such project- or programme-specific data collection serves the long-term interest of developing the national capability is doubtful. But its growth indicates the demand for certain data and the perception of many users that the official statistics agencies cannot be relied on to meet this demand.

And lastly, many small-scale, but often valuable, surveys are carried out by universities, research institutes, and individuals. These can complement and supplement the official data collection effort if there is regular communication and dialogue between the bodies concerned. Often this has not been the case: regrettably, official statisticians have been somewhat intolerant of academic case studies and academics over-cautious in keeping their distance from official 'contamination'.

Three major requirements seem to us to be essential if data collection is to develop in a sustainable manner in the near future. These requirements will provide the focus of our presentation in later chapters and are introduced briefly here.

(a) Simpler, cost-effective methods of data collection. As mentioned above, progress has been made in some areas such as demography and nutritional status. But with household economic and agricultural variables the enumeration methods are too demanding in time and resources, leading to expensive surveys that are not reported in time to meet user requirements. Governments are rightly chary of supporting such ventures on a continuous basis. Statisticians and surveyors must find ways of overcoming these constraints if their credibility is to be established.

(b) More efficient and effective use of sampling methods. The widespread use of large, inefficient sampling designs is partly due to the methodological problem of data collection stated under (a). Excessively clustered samples with resident enumerators are forced on the sample designer by the need for objective measurements and repeated interviews, and the lack of transport facilities. But the price has been too high. In addition, the ambition to provide estimates for small administrative areas has led to sample sizes that outstrip the ability to manage the field operations. For many purposes quality data from small samples will meet the main survey objectives. Further, the important role of case studies is often overlooked. (c) Greater emphasis on providing timely information to primary users. Data collection is not an end in itself: unless the data can be processed, analysed, and converted into information in a format that can be assimilated by the users the effort is both pointless and costly. One of the major problems that arises when a continuous survey programme is established is the preoccupation of the limited professional manpower with the maintenance of the data collection effort at the expense of information production. Allied with this requirement is the need for improved collaboration among all those involved in data collection so as to achieve the efficient blend of various disciplines and the avoidance of duplicated efforts which often result in conflicting estimates.

Notes

1. Seng, Y..P., 'Historical Survey of the Development of Sampling Theories and Practice', *JRSS* (A) 114, 1951, p. 217.
2. Ibid., p. 221.
3. Ibid., p. 222.
4. Neyman, J., 'On the Two Different Aspects of the Representation Method: the Method of Stratified Sampling and the Method of Purposive Selection', *JRSS* 97, 1934.
5. Hansen, M. H., and Madow, W. G., 'Some Important Events in the Historical Development of Sample Surveys' in Owen, D. B. (ed.), *On the History of Statistics and Probability*, M. Dekker, Inc., New York, 1976, pp. 87–8.
6. Mahalanobis, P. C., 'On Large Scale Sample Surveys', *Phil. Trans. Roy. Soc.*, Series B, 231 B 584, 1944 (communicated in 1943); and 'Recent Experiments in Statistical Sampling in the Indian Statistical Institute', *JRSS* 109, 1946.
7. Mahalanobis, 'Recent Experiments', pp. 332–3.
8. Hansen, M. H., and Hurwitz, W. N., 'Relative Efficiencies of Various Sampling Units in Population Inquiries', *JASA* 37, 1942.
9. Yates, F., *Sampling Methods for Censuses and Surveys*, Charles Griffin, London, 1981.
10. Deming, W. E., *Some Theory of Sampling*, Wiley, 1950.
11. World Fertility Survey, Proceedings of World Fertility Survey Conference, London, 1980.
12. FAO, *Programme for the 1980 World Census of Agriculture*, Rome, 1976.
13. Mason, J. *et al.*, *Nutritional Surveillance*, World Health Organization, Geneva, 1984.
14. Casley, D. J., and Lury, D. A., *Monitoring and Evaluation of Agriculture and Rural Development Projects*, Johns Hopkins Univ. Press, Baltimore, 1981.

2

Deciding What Data to Collect

He appears to have a passion for definite and exact knowledge.

A Study in Scarlet

2.1 ASSESSMENT OF PRIORITIES

The collection of data in developing countries is almost always an expensive operation. The greater relative need for physical counts and measurements by the enumerator adds substantially to the cost per sample unit. Surveyors cannot afford to waste scarce resources on collecting data that will not be put to good use. Certainly it is no defence to state that the surveyor's job is to collect and process data and not to worry about their use. Many officials in government, banks, and research institutions are ready to suggest a never-ending sequence of data requirements. Few of these potential data users will have considered the cost of obtaining the information against its potential value; indeed, many will have neither the knowledge nor experience to enable them to make this assessment. The surveyor should be able to do so. He must ensure that he takes responsibility for surveys only when there has been a conscious decision that the resulting data are important, will be used, and that equivalent information cannot be obtained in a cheaper way. Inevitably this process will involve assigning priorities to data sets within the programme for which the surveyor is responsible; and this means, of course, that some users must accept a lower place in the total programme than they would like.

In discussing how the surveyor should deal with priorities we shall consider the various steps in sequence. It must be remembered, however, that many of the problems are interrelated and they should not be dealt with in isolation. As a first round, provisional alternatives are delineated and their implications considered; the whole process is then repeated until what appears to be the best package of interrelated decisions is achieved. The key question throughout the whole planning, and particularly at the very early stages of the survey design, is 'Why are we collecting these data?' This question relates to the topic of the survey, to the function the main (or primary) users expect the results to serve, to the range of its content, and to each individual question in the questionnaire.

The surveyor's view of priorities, and how they are to be determined, depends on his place in the organization and whether he is employed in government, a research organization, or individual research. We wish to argue that workers in the last two situations, if they are working in developing countries, should consciously pay greater attention to fitting in with general national (including short-term) needs than they might if they were working in a more developed country. We should make plain at the outset that we are not suggesting any curtailment of scientific or academic independence. The research worker must retain his professional standards in this respect; only in this way can he make his particular contribution. However, he can make an especial effort to see how his work can meet some of the community's needs, and how it might fit into official programmes. If his survey has little value to the country or to the people who are the subjects of it, he should not be surprised if requests for special assistance or tabulations are received by official statisticians with some lack of enthusiasm. Not only do they have many other things to do: they have usually had disillusioning experience of irrelevant, and often ineffectual, surveys carried out as individual research projects by workers who often leave the country before making available locally such results as they have.

If the survey as broadly conceived passes the test of usefulness and is placed high on the priority list, the next stage is to keep a watchful eye on the selection of the specific issues to be included. There is a natural but unfortunate tendency to start expanding the original objectives, and to include questionnaires or modules that are not required to achieve them. This is the 'as we're there, we may as well ask them this' school of thought. For each section of the survey the test should be the same, 'Why do we need these data?' The answer may be that additional data are needed by someone for something, but not to achieve the original objectives. There will be occasions when one survey can justifiably carry another 'piggyback': but two questions must be answered before adopting such a procedure—'Are these "secondary data" important enough to justify their inclusion?' and 'Are the chances of collecting successfully the primary data lessened by including these questions on other topics?' We return to this issue when considering different types of surveys in succeeding chapters.

At the last stage of deciding the survey content, the inclusion of each individual question must also be subjected to the same test: 'Are the data needed?' In many surveys questions or variables are included which will not be analysed or serve for identification or cross-checking purposes. In listing the household members in a household survey there is a tendency to ask each person his age, education, tribe, religion, etc., even if the objectives of the survey and the tabulation plans do not require some or all of these items of information for everyone. Education may be related to food consumption if we are studying a heterogeneously educated population, but the relationship is likely to be a complex one with many other variables playing a role. Before including this aspect we must be sure it is intended to tabulate by education.

Two attitudes may unduly dilute the adoption of this sequence of strict limitations. The first is a desire to 'play safe'. It is easy to think of variables which may be relevant to the main objectives of the survey: it requires confidence to reject in advance a proposed addition that is vaguely connected with the survey topics, and which *could* prove to be important. Most experienced surveyors can recall occasions when they overlooked a point which turned out to be important for some aspect of the final report. The second attitude arises from a desire, praiseworthy in itself, to include questions on related subjects of interest to users. This wish is likely to weigh heavily with government statisticians since they generally have to satisfy the demands of many users and are used to reviewing their programmes in this light, standing in themselves as a proxy for these users. Appropriate action varies of course from case to case; but in view of the pressures to expand the topics to be covered in any survey, we recommend adherence to the basic aims of the inquiry.

2.2   USER–SURVEYOR DIALOGUE

The need for the surveyor and the data user to talk to each other is a universally accepted maxim; but like many other such maxims it is often ignored. As part of a drive to develop a survey capability, a statistics office may wish to prepare a five-year programme. Extensive dialogue takes place with statisticians and surveyors belonging to technical assistance and specialized agencies so that the sampling plan, survey methodology, and variable definitions meet international standards. But in some actual instances there is very little dialogue with the potential users of the data. It is assumed that labour force data and child morbidity data, to take two examples, are important to national agencies and planners: this may be so, but what precise data are most needed, in what format, and at what time are questions often unaddressed.

Too few surveys are initiated by, and focused precisely on, well-identified requirements of a particular user: too many are launched on the basis of a consensus amongst the surveyors as to what data the user is likely to require rather than on the knowledge, gained from extensive dialogue, of the actual information gaps troubling the user at the time.

A particular case is when the survey is organized and executed by the user agency itself, for example a survey in an individual research project. The necessary dialogue should still occur, even if it is a case of the research worker putting on two hats and talking to himself.

The way in which the dialogue proceeds will be dependent on the type of survey, the organization commissioning it, and the one carrying it out. An *ad hoc* survey mounted by a research institution as part of a funded project will normally have a narrower range of users to consult than a government surveyor mounting a continuous multi-subject survey. The following sections set out some general suggestions, which will, of course, have to be adapted for application to different types of inquiries.

2.3 DEFINITION OF THE PROBLEM AND THE SURVEY OBJECTIVES

We concentrate first on the debate with the commissioner of the survey, or—when several interests are concerned—the main customer. The surveyor should take great care at this stage, since these first discussions will set the whole tone of the survey and probably determine its effectiveness. It is useful if the surveyor has knowledge of the field of enquiry so that he can probe the customer's requirements. However, even without such knowledge he can get quite a long way by using standard probes, such as 'What do you mean by 'X'?' 'How do you think 'Y' may be related to 'X'?' 'Why should information about 'Z' help you to deal with your problem?'

Many surveys are requested because it is hoped that the results will help those commissioning the survey to take policy decisions about a problem they are facing. The first result of the user–surveyor dialogue is usually to make the customer define his problem more precisely. In this early stage of the dialogue, the customer may even get a little resentful when the surveyor cross-examines him in order to set the limits of the survey. The question 'What do you mean by unemployed?' may evoke the response 'It's obvious, everyone knows.' The customer, partly because of his involvement with the problem, thinks that certain aspects of the subject are self-evident. Further, since he has to frame administrative remedies, he is looking for a report structure that parcels the situation into neat boxes, often hoping for a simple dichotomy of Yes/No that does not match the complexity of the reality the survey has to reproduce.

The interrelationship between policy options, existing information, and new data is set out in the context of nutritional surveillance by an Expert Committee.

At the outset a dilemma has to be faced: on the one hand, it is impossible to develop an efficient system of collecting information without knowledge of the purposes for which the information is to be used. On the other hand, problems cannot be defined and policies formulated in the absence of information. Initially, any programme of surveillance and any definition of policy can openly be based on the information that is available and on the objectives that appear to be important at that time. However, surveillance and any definition of policy can only be based on the information that is the system itself, and lead to redefinition of objectives.[1]

It will usually require a great deal of effort to agree upon a common approach. If the surveyor knows little about the substantive field in which the inquiry is to take place, there will normally have to be at least two meetings, and the surveyor will have to familiarize himself with the customer's problem in the period in between. Eventually, it must be possible to establish whether the survey method will be useful to the customer, and, if so, to define the main lines of the inquiry.

Naturally, the issues will vary from survey to survey, but two obviously fundamental aspects will recur. They are: (a) time: the period to which the inquiry refers and the period in which it will take place. These are usually not independent, since most surveys of human population involve memory; and (b) place: is the inquiry to be national, regional, or of a smaller area? Is it rural and/or urban? Has the most appropriate area been properly identified already? for example, areas with populations suffering from high malnutrition or the project impact area in an evaluation survey.

Demographic characteristics will also recur. Characteristics such as age and sex are almost universally required. Others relating to household structure may be essential for the inquiry, or may just be suggested for identification or as explanatory background. (As a general rule, such general purpose sections need to be restricted to basic characteristics, otherwise they tend to dominate the questionnaire unduly.)

The selection of the topics that are to be dealt with in the survey will also set limits to the types of survey that are appropriate. In addition to setting times and places, the methods of inquiry and level of enumeration will have to be matched with the agreed objectives. Clarity and precision at this stage will prevent waste of time at later stages of planning when, for example, the objectives are crystallized in detail and embodied in specific questions.

2.4 EXISTING KNOWLEDGE

The surveyor will often know, at least in broad outline, the scope of the information relevant to the subject of the survey that is already available. Nevertheless, he should explore the situation with the user, since the latter may know local or international sources better than the surveyor, or may have contacts to whom the latter can be directed. Secondary sources may be used in four ways: (a) as an alternative to the survey, or part thereof; (b) as independent additional information; (c) as a check on possible survey biases; (d) as a means of improving the survey estimates.

The following example is based on a survey with which we are familiar. The background to the survey is that credit, either in the form of cash or agricultural inputs, had been given to small farmers who were members of co-operatives. The survey objective was to discover how much of this credit had actually been used for the purpose for which it was granted, that is, to produce food crops. A sample of members was selected from each of a sample of participating farmer co-operatives.

The amount of credit each respondent received is needed, but his memory may be faulty. Suppose records of this information are available in the co-operative office. This source can be used as an alternative to asking the respondent, or as a check on his reply. (It can also be used for the purpose of drawing a stratified sample, but this aspect is not within the context of the present discussion.)

The respondents may report the quantity of fertilizer received but may not know the chemical composition of this fertilizer. The co-operative records may not state how much fertilizer each respondent received but may show its composition. This will be a useful piece of additional information. The respondents may be suspicious of the purpose of the survey despite attempts at explanation. They may understate, therefore, the credit they have received, thus biasing the survey. Some check of this can be made if the total amount of credit issued to all co-operative members is available. Or it may be known that there was a minimum credit package, so that respondents' replies indicating receipts less than this suggest bias.

The data collected from the respondents include the value of credit applied to food crops, and the total value of credit received. Estimates of the total value of credit used on food crops may be improved by the use of ratio or regression methods if the total value of credit issued is accurately known from co-operative records. This use of supplementary information to improve estimates obtained from small samples is described in most standard texts on sampling.

We wish to emphasize that surveyors should be alive to possible improvements which may be obtained by exploiting secondary or alternative sources of data. It is well worth making extensive inquiries to find them and to check on work in progress. There is nothing more galling than to mount an elaborate survey only to find at the publication stage that the information is already available. It has happened in the past and can easily occur again since government bureaux, institutions, and individual research projects are usually inadequately co-ordinated.

The validity of the secondary sources must of course be carefully assessed. Unfortunately, statistics—especially official statistics—may appear to be more accurate than they are. In many countries the absence of a survey capability results in data being recorded by officials whose responsibilities lie elsewhere; and whose objectivity may be affected when it comes to reporting situations for which they themselves are partly responsible. Agricultural extension agents are not, as a rule, a very reliable source of crop area and yield data, although, properly sifted, their information may be useful as a secondary source to supplement more objective data.

Their information may also be valuable for crops grown in small, local areas, which may not be adequately covered in a national sample. These crops may be of greater economic importance than the area they occupy suggests. Often they are crops to which the department of agriculture is paying great attention, and extension agents may be able to give quite reasonable estimates as they know every farmer in their area growing them. These data may well improve the crop tabulations derived from the survey. Discussions with middle-level officials on such matters should be seen as part of the user–surveyor dialogue.

Medical and nutritional data may be available from records maintained in hospitals and clinics. It is sometimes possible to put these data to use in a more general way; but it requires great expertise, for such data are seriously biased if applied to the general population. The same comment applies to data of crop yields or livestock carcass weights, etc., which may be based on research station data, and so scarcely reflect the situation in the country as a whole.

2.5 SECONDARY USERS

Following up related sources and recommended contacts will make the surveyor aware of other people and institutions who are likely to be interested in the results of the proposed survey. Some of them may give him useful insights that help in the continuation of the primary user–surveyor dialogue. Others may urge the surveyor to extend his survey or to allow them to add 'just two or three questions'. Whilst it would be too sweeping to advise that such requests should *never* be attended to, we fear, as already stated, that many surveyors are too ready to expand the objectives of the primary user in order to meet secondary requests that may appear (often misleadingly) to require only a slight modification of the investigation.

Sometimes it may be difficult—and occasionally even impossible—to identify a single main user, particularly of government statistics. When, say, an *ad hoc* survey of agriculture is being considered by a government, there may be no single user that the survey designer can regard as paramount. The department of agriculture, the agricultural research institutes, agricultural economists in university or development institutes, and development planners in the planning ministries are only a few of the potential users of the data. The needs of some conflict, from a survey design point of view, with the needs of others. However, if the survey has found its way on to an integrated government data collection programme, the question, 'Why are we collecting these data?' should already have been addressed and answered. The answer might be that the objective is to provide national aggregate data on the composition and structure of agricultural holdings. If so, the primary users may be the ministry of agriculture and other macro-planners in the agricultural sector. In this situation, agricultural research workers are secondary users who are not being catered for: the dialogue with them may be limited, and must make the situation clear. One of the main pitfalls at the survey design stage is to attempt to be all things to all men, and so end by satisfying none. It is part of the surveyor's lot, after the results of a survey are published, to be doomed to listen to a chorus of 'It's a pity you didn't ask . . .' or 'If only the data were classified by . . .' . These recriminations may be justified in some cases, but in many others no report at all would have been published if the surveyor had broadened his inquiry. The surveyor has failed, however, if the main needs of his primary users are not met. The failure is even more inexcusable if it occurs because he has widened and diluted the objectives to satisfy other, perhaps vaguely defined, uses, and so hindered the collection of the 'primary' data.

2.6 MORE DETAILED EXAMINATION OF OBJECTIVES

After the general objectives of the survey have been confirmed, it is necessary to discuss and agree with the user about a more detailed specification of the issues that are to be investigated. The response to the question, 'What precisely do you want to know about it?' may well be in effect, 'Everything'. A common reason for such an amorphous answer is that the customer feels he cannot judge whether some variable is relevant until he 'sees the figures'. Frequently, when even the general dimensions of a problem are not clear, there may be an opinion that no policy alternatives can be explored until information is obtained; or, in the most desperate case, the user may hope that the survey results will themselves indicate a policy.

The surveyor must, in problem–policy-oriented surveys, press his customer to think in terms that relate possible alternative remedies to the specific data to be collected. This will not only make the final survey more effective: it will also make the policy maker think hard about his problem in a survey context. The surveyor following this process may come up against an unfavourable response; a reaction suggesting he is going beyond his proper sphere. His questions, 'Why do you want to know this?' and 'What use will you make of the data obtained in this section?' may get the reply, 'What business is it of yours?' Nevertheless, he should not be put off. He should explain his reasons, and try to convince his customer that it is in both their interests to proceed in this fashion.

The listing of the items to be included and their definition should be discussed with the user. When his attention is directed to this level, much of the user's initial vagueness may disappear. For example, the user may want income by a particular occupational grouping. The surveyor may be able to point out that many statistics in related fields classify occupations in a way other than that proposed. The user can then decide whether to change his first proposal, and adopt the standard classifications, or to agree on the collection of more details so that both classifications can be presented.

In certain technical surveys, some of the concepts and definitions proposed by users may be too difficult for an enumerator to follow. The definition by an animal breeder of an 'improved' animal may require an examination of the animal that the enumerator has neither the time nor the training to carry through. The definitions agreed upon must be acceptable to the users, but must also be capable of being easily applied in the field by someone untrained in the particular discipline under study. We know of a survey that required the enumerator to classify animal deaths by cause from the vague and general description volunteered by the farmer. Needless to say, the attempt was unsuccessful.

Neither is it proper to use a methodology and concept that, even if understood, should not be applied. If general purpose enumerators are to be used, questions and observations requiring clinical judgements should not be included. Even if the enumerator can be trained to take the observations, questions of ethics still arise. Medical personnel may be needed in such a case. The surveyor should not presume on the goodwill of a population just because they are unaccustomed to voicing complaints. The test is whether the approach to be taken is acceptable to a respondent who is able to articulate opposition and has the social position to do so. If not, it should not be applied to any respondent without very careful consideration.

Once the surveyor has reached agreement with the primary users on the basic contents, concepts, and definitions, he and his associates should get on with the job of designing the survey and the questionnaires, and accept the responsibility that is now rightfully his. It is at this stage that things often go wrong. Draft questionnaires are circulated not only to the users with whom discussions have been held, but to a much wider group. Comments are requested and additions solicited. Committees are formed to vet the drafts. All this is an abrogation of responsibility by the surveyor. Questionnaire design, that is, deciding how to translate detailed data needs into specific questions, is a highly technical, skilled exercise. Most users, and many non-survey statisticians, have little idea of how to go about it. The survey designer who trails his drafts around the corridors is likely to end up with a host of comments that, if adopted, will weaken the focus of the survey, and prevent his meeting his main objectives. Excessive activity of this kind usually occurs either because the surveyor has failed to engage in the appropriate levels of user–surveyor dialogue *before* designing and drafting the survey and questionnaires, or because he lacks confi-

Data Collection

dence. The users should not be expected to do this work. The unhappy fact is that many survey designers are insufficiently skilled in this stage of their work. Some find it daunting, some tedious, and others, who are absorbed with the mathematics of sampling, find it of less interest.

This emphasis on the surveyor's responsibility should not prevent him from consulting a colleague who may be more experienced in one phase of design than himself. Surveyor–statistician dialogue is as important as user–surveyor dialogue. Further, there may be one or two potential users or experts in the subject area who can be usefully consulted. The action we are pillorying is the widespread, almost indiscriminate, circulation of drafts for comment.

When the questionnaire is ready, the surveyor should visit the primary users and explain precisely what he thinks will be achieved, and to what extent the original objectives will be met. The users may then wish to take advantage of the opportunity for second thoughts, and slightly change the direction or the content of the survey. This will require redrafting by the surveyor, but this is a necessary part of the preparatory process.

In far too many cases, the surveyor is not involved as early as he should be. Especial care needs to be taken when a user comes with a design and questionnaire already prepared. In order to save time, and also sometimes to avoid upsetting the user, the surveyor may merely tinker with the problems. He will then be in danger of being saddled with responsibility for faults in a survey that were already present when it was first brought to his attention. The better course is to persuade the user to return to the beginning, and allow the surveyor to exercise his rightful role. A worse, but still common, situation is that a statistician is not called in until a much later stage and is faced with data already collected, and sometimes even partially analysed. He is then expected to meet the usually impossible demand of matching the results of an ill-designed inquiry to the user's objectives.

2.7 RESOURCES AND TIME

The surveyor must discuss with the user the interrelated questions of the resources available for the survey and the time the survey will take. The framework within which this discussion will take place varies. If the survey is given priority and is well funded by the main user, the resource and timing constraints may be minimal. On the other hand, an official agency wanting a survey that has to be fitted into a round of a national sample may have to join a queue.

Certain time constraints may arise from the inherent nature of the survey topic. Seasonal variations may mean that a full year has to be covered before any useful conclusions can be reached. Some technical surveys require an extended period of training for enumerators and supervisors, and it may be necessary to wait for the arrival of special equipment. As a general rule, users tend to have unrealistic expectations about the time and money required for a good survey, and many will be anxiously seeking short cuts. If the short cut proposed involves reducing the survey objectives, it should obviously be explored: if it requires the reduction of training and transport, or interferes with other essential processes, it must be resisted. The surveyor will be blamed for anything that goes wrong, even if the initiating cause of the survey failure is the parsimony of the user.

One common trade-off is that between resources and the level of disaggregation. The surveyor should inform the user of the levels of disaggregation that will be possible from a given effort. Discussion regarding inter-district comparisons is irrelevant if the surveyor already knows that the resources proposed will permit only a sample size sufficient to produce national or regional estimates. It is irrelevant, that is, in terms of detail: the question whether regional estimates are any good for the user's purposes is of course highly important. If district estimates are indispensable the surveyor must demand the appropriate resources. What, obviously, he must not do, is to appear to indicate that the user's needs, say, for district figures, will be met, when in the event they cannot be. Unfortunately, this issue is often 'fudged', sometimes because of over-optimism and a desire to please. Many governments have a policy of decentralized planning at a district level. But the resources required to collect a range of statistics using sample surveys are usually not available. The level of disaggregation needs attention by topic. Given the sample size, is it possible to tabulate income by occupation of head of household, or by family size? Can the manpower needs of industry be reported on by ISIC (International Standard Industrial Classification) groups?

Most textbooks dealing with sample surveys include sections on calculating the sample size needed to achieve stated objectives, and on choosing the design that will minimize costs for a given level of accuracy. In this affluent context the surveyor's job is to listen to the user enumerate his requirements and then make the necessary calculations and choose a sample design. Unfortunately, reality rarely conforms to this happy ideal. The more usual situation is that the surveyor is aware that he cannot possibly obtain resources, or handle the administration, for a survey above a certain size, which is below the size that would be calculated using the orthodox techniques. The question is, therefore, 'What results can I usefully achieve given the resources?' rather than 'What resources do I need to achieve this objective?' Private researchers are often in the same situation. If they are setting out on their first survey they may easily overestimate what they themselves can do by 'working a little harder'.

It is obviously essential that there be no misunderstanding between the user and surveyor. They must record their agreement about the resources that are to be made available, the times that the data and reports will be produced, the level of disaggregation that will be provided for, and the range of error to be expected.

Many statisticians find that the initial request they receive for advice is couched in the terms 'How big a sample should I take?' This question cannot be answered, except in a form beginning 'It depends . . .'. Nevertheless, the approach even in this naïve form enables the statistician to initiate a dialogue that eventually encompasses the whole design.

2.8 TABULATION AND ANALYSIS

Consideration of tabulation and analysis requirements is often postponed until the survey is well under way, sometimes on the ground that no sensible decisions are possible until the results of the field work begin to come in. It is, of course, very likely that some information will come to hand during the survey that will suggest features or relationships that were not foreseen, and which could not have been provided for in advance. But this likelihood should not prevent the preparation of at least a partial tabulation programme. Indeed, such a programme is an essential part of the iterative formulation of an effective survey plan.

In the first place, it helps to proceed with the detailed examination of objectives, discussed in Section 2.6 above. One of the best ways of clarifying ambiguities and vagaries is to consider what variables are to be cross-tabulated, and the major categories or class intervals required. A number of draft tables should be prepared showing the form in which results are required—an activity sometimes referred to as 'stub analysis'.

Existing information should be used to estimate—even in a very crude fashion—the relative magnitudes expected in the cells of the proposed tables. It will often be impossible to do very much, perhaps little more than identifying those cells in which the larger magnitudes are expected. Sometimes, however, it may be possible to fill in roughly the cells in the marginal (subtotal) row and column of the draft table, and a consideration of these figures may throw light on what is to be expected in the body of the table. Even if the results of such preliminary working provide only broad orders of magnitude, they will still contribute to the selection of the most appropriate design for the survey.

Further, the allocation of resources discussed in the previous section should take into account the time and cost of analysis. These are generally underestimated. The major problem in survey design is to obtain a proper balance between the demands of different stages of the total process. It is usually well known that field work will be difficult and costly, and the problems there obviously occur at an earlier stage than those arising in analysis. This 'visibility' diverts attention from the needs of subsequent steps in the survey, often unbalancing the whole operation.

2.9   RAPID ASSESSMENT

In many user–surveyor dialogues, the point is reached when the user asks the surveyor to produce quick results to help meet a particular deadline or cope with an immediate crisis. The request may be for the surveyor to produce something relevant from existing data, without or before a survey; to produce preliminary survey results before the analysis of the survey is completed; or to produce more detailed analyses than have been provided for. But most commonly the surveyor is urged to produce results by faster, informal methods. Never mind the need to establish frames from which to sample, or the care needed to pilot questionnaires; the surveyor is urged to go out and get a 'quick fix' on the situation. In recent years a school has grown up that emphasizes the need for rapid informal approaches to the assessment of rural development and the problems of such development.[2] We comment on informal assessment methods in Chapter 5, but a general word on surveyor responsiveness is in order here. The surveyor's natural reaction should be to help the user, but he must be careful. He may attempt to satisfy the user's wishes, subject to the user's accepting that there are all kinds of reservations that have to be made about the 'quick and dirty' results produced. Later, if policy decisions taken as a consequence turn out to be wrong, the surveyor's reservations may well be forgotten and he will find he is landed with the final responsibility for the defects of policy; and he will have to put up with the adverse effects on his, and the profession's, reputation.

There is a real dilemma here. The extension of statistical work and the growing acceptance of the need to involve the surveyor more regularly and immediately in policy issues can be set back if the surveyor behaves as if he can operate only in ideal conditions and cannot adapt to the stresses and strains of helping decision takers effectively. Statisticians and surveyors are to blame in that they have frequently proposed data collection methods and timetables that will not produce the information within the policy-dictated deadlines. If users, and some practitioners, are turning to inappropriate methods it is because they have not been informed of methods that are rapid and cheap, but which are within the range of valid procedures.

Notes

1. Joint FAO/UNICEF/WHO Expert Committee, *Methodology of Nutritional Surveillance*, WHO, Geneva, 1976, p. 7.
2. Chambers, R., *Rural Development: Putting the Last First*, Longman, London, 1983.

3

Censuses

If we could fly out of that window hand in hand, hover over this
great city, gently remove the roofs, and peep in at the queer
things that are going on . . .

A Case of Identity

3.1 INTRODUCTION

Failures to obtain response and errors of measurements and processing arise in both sample surveys and censuses. Comments in other chapters about these matters, therefore, also apply to censuses. This chapter deals with censuses separately for two reasons: first to consider some special aspects of census taking; and then, in the course of doing this, to assist those who have never been involved in a census to understand some of the difficulties encountered. This understanding should enable them to make a better assessment and use of census data. We shall deal first with the population census, then briefly discuss other forms of censuses including those conducted by mail inquiry.

3.2 POPULATION CENSUSES

The advantages of sample methods of inquiry have already been set out; a brief account of reasons for taking censuses is now appropriate. The choice is not just between a census or a sample; a number of countries now combine in one operation a census of a number of characteristics with a sample covering other items.

The most important reason for a census is the political and administrative requirement for separate figures of the main features of population for each of the smallest administrative units used by government. In this form the census provides a national inventory, a picture of the situation existing at the time of the census. The value of information from a single census is, of course, enhanced if it becomes part of a series arising from a programme of censuses at regular intervals.

A population census is often combined with a housing census. It is usually preceded by the construction or updating of maps showing small area boundaries over the whole country and results in household lists. These maps and lists can be used as frames for subsequent surveys. Census data can be used to improve survey design, for example, as a means of stratification. They can be of value as supplementary information to improve sample estimates. A substantial source of data in developing countries is provided by the returns and reports of district and provincial officers, particularly technical and professional officers in sectors such as agriculture, education, and health. A basic inventory of the nation provides a background against which the relative value of this uncoordinated network of information can be judged.

A population census, even one that uses a short and simple questionnaire, is a major operation. The accompanying chart, calculated from percentage factors that appear reasonable in the light of available information, shows why this is so. Every million inhabitants require up to 2,200 field staff to enumerate them, and contain about 220,000 people from whom these staff could be chosen.

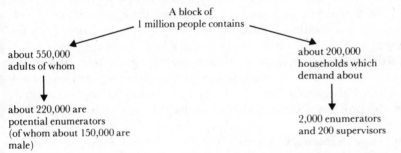

A block of
1 million people contains

about 550,000
adults of whom

about 220,000 are
potential enumerators
(of whom about 150,000 are
male)

about 200,000
households which
demand about

2,000 enumerators
and 200 supervisors

Thus 1 per cent of the potential enumerators and supervisors will be involved, and be involved for at least three or four weeks if the training of staff, the enumeration, and the checking and dispatch of forms to the centre are to be of an adequate standard. In some countries only males have undertaken the role of enumerator; the proportion of actual to potential will be higher where this restriction still applies.

The recruitment and training of the supervisors and enumerators are most demanding exercises because of the numbers involved. Training at several levels will be inevitable; the census designers train district census officers, who train the supervisors, who in turn train the enumerators. This sequential process leads to a grave risk of distortion of instructions as they are passed down the pyramid—a risk that is accentuated by the fact that most of those involved in the later stages of training are themselves inexperienced in data collection.

Difficulties in census taking may be illustrated by examples from Western Asia.

The widespread illiteracy in Arab countries is one of the main obstacles to obtaining accurate data. ... In many instances it is difficult for the researcher to enter the household. In certain cases, it happens that the researcher did not actually see any of the household members. He obtained the data from one of the women behind a closed door. . . . Multiple nationality proved to be most common among nomadic bedouins who cross national borders. The problem is still uncontrollable as most of them have more than one nationality and they do not usually show their papers to enumerators. . . . Certain house-owners of limited income who got their houses through the State were trying to deny sharing the house with other tenants.[1]

There is great disparity in the country's climatic conditions, and this made it difficult to determine the time period for taking the population census. For almost every region there was a specific time of the year that was suitable for the census operation there but not elsewhere. February was the only month climatically suitable . . . in most parts . . . since the month selected did not fall within a school vacation period, it was necessary to close the schools and the university through the census period.[2]

Most developing countries decide that the cost of a full population census has to be accepted, and can generate sufficient enthusiasm and interest on a national scale for it to be carried through satisfactorily. It requires the mobilization of resources within government, particularly the provincial and district administration, the education system, and urban authorities. Planning will need to start two to three years before the enumeration date, and the analysis may well not be completed until three to four years afterwards (see Figure 3.1 for a specimen census calendar, which also provides a useful check-list of the sequence of operations to be followed).

The basic framework depends on the administrative structure of government. The primary analyses will be produced for the smallest units and then aggregated into larger groups according to administrative areas and to any special planning requirements. Although, within the framework, the statistical organization will play a large part in determining the form of the census, the quality of the information collected will depend very much on the district administration. The character of this administration and the readiness of its local officers to respond to the centre and to accept direction on various issues is, therefore, crucial. One of the essential prerequisites for a successful census is the development of good relations between the statisticians and the administrators; the selection of the right officer to be in charge of the administrative aspects of the census is one of the keys to success. He must be highly regarded by his fellow administrators, since he will often have to check and correct them without losing their support or reducing their commitment to the census.

In many countries administrators have close links with the local political structure; there is, therefore, a danger that political considerations may affect the way in which the census is carried out. There are often considerable regional disparities in modernization and education which are likely to affect the relative availability of resources for local enumeration in local languages, and the readiness and ability of the local inhabitants to respond. The census organizers at the centre need to keep a close watch on the way different regions operate in the pre-census period, so that they can judge how to dispose of their resources to the best advantage during the actual enumeration.

The census programme may be disrupted, and sometimes the census date changed, owing to political causes or natural disasters. For example, in the 1971 Indonesian Census:

The training of enumerators was originally planned to take about three months, using only a few instructors who would travel from area to area to conduct the training programmes. This procedure would have helped to ensure the quality and consistency of training. However, a government regulation prevented meetings in the three months prior to the July, 1971, election and the training schedule had to be compressed mainly into the month of August. Three times as many first-stage instructors as originally planned had to be employed, with a consequent increase in the variability of instruction and a probable adverse effect on the quality of the final enumeration.[3]

The census is said to be characterized by: (a) individual enumeration; (b) universality within a defined territory; (c) simultaneity; and (d) regular periodicity. Some censuses in the past did not attempt (a), but it is standard practice nowadays. The second criterion is not quite so limiting as might appear, since the territory can be defined to exclude inaccessible areas, or those parts of the country where the administrative writ does not run. In these cases the territory enumerated will be defined, but it will be less than the complete national area. Even within the defined area, however, there may be groups who are difficult to cover, such as nomads, or homeless people in urban areas. The criterion of simultaneity is important. The enumeration should be conducted within a very short and well-defined period of time to reduce omissions and duplications and to maintain the impetus of the operation. This requirement, together with the scale of the enumeration, makes a census unique in its demands for a very large, and temporary, enumerator and supervisor force. The periodicity recommended in most census manuals is ten years: an increasing number of countries are approaching a programme of this kind.

Another feature that we wish to single out in this brief review is the importance of maps. An increasing number of countries now have maps produced from a basic geographic survey, but each census needs to be accompanied by a review and updating of them.

During the enumeration particular attention should be paid to ensuring that, if ambiguities remain on the maps of enumeration areas, the enumerators together do cover the area without omission or overlap. Because *every* area is to be covered, enumerators in contiguous areas need to be co-ordinated by the supervisor so that both are using the same boundary even if there is some doubt as to its correctness. Thus, if the boundary, shown as a footpath, is ambiguous because there are two

1. FRAMING OF CENSUS ACT AND CREATION OF CENSUS COMMISSIONER'S POST--
2. OBTAINING USERS VIEWS ON CENSUS QUESTIONNAIRES - - - - - - - - - - - -.
3. FORMATION OF ADVISORY COMMITTEE -
4. RECRUITMENT OF KEY PERSONNEL (HEADQUARTERS DATA PROCESSING - - - - - CENTRE AND ZONAL OFFICES)
5. PROCUREMENT OF ACCOMMODATION, TRANSPORTS AND FURNITURE FOR HEADQUARTERS
6. FORMULATION OF CENSUS QUESTIONNAIRES - - - - - - - - - - - - - - - -
7. OBTAINING ADVISORY SERVICES AND PRETESTING OF CENSUS QUESTIONNAIRES - - - -
8. FRAMING OF CENSUS RULES -
9. FINALIZATION OF CENSUS QUESTIONNAIRS, TABULATION SCHEME, CLASSIFICATION - OF INDUSTRIES AND OCCUPATION
10. PRINTING PLAN AND PREPARATION OF MANUAL OF INSTRUCTIONS FOR SUPER- VISOR AND ENUMERATORS.
11. RIVISED ESTIMATES (1972-73) AND BUDGET ESTIMATES (1973-74) - - - - - - -
12. PROCUREMENT OF ACCOMMODATION, TRANSPORT AND FURNITURE FOR ZONAL OFFICES -
13. TRAINING PROGRAMME AND PREPARATION OF TRAINING MATERIALS - - - - - - --
14. ESTABLISHMENT OF HAND SORTING CENTRE - - - - - - - - - - - - - - - - -
15. ESTABLISHMENT OF DATA PROCESSING CENTRE (ACCOMMODATION, FURNITURE- AND AIR-CONDITIONING)
16. PRINTING OF MANUAL OF INSTRUCTIONS QUESTIONNAIRES, HOUSING SCHEDULES,- CONTROL FORM ETC.
17. PREPARATION OF PUBLICITY PROGRAMME AND PRINTING OF PUBLICITY MATERIAL
18. PROCUREMENT OF CENSUS PORTFOLIOS, BALL-POINTS, PENS, INSCRIBERS, MAG- - NETIC TAPES, COMPUTOR, STATIONERY ETC
19. APPOINTMENT OF C.D.O. CHARGE SUPDT. CIRCLE SUPERVISORS / ENUMERATORS - - - -
20. FINALIZATION OF CODE GUIDE AND PRINTING THEREOF - - - - - - - - - - - -
21. COLLECTION, TRACING, UP-DATING, RETRACING, AND PRINTING OF URBAN AREA MAPS --
22. COLLECTION OF VILLAGE LIST, UP-DATING, PREPARATION OF SKETCH MAPS AND- DELIMITATION OF RURAL AREA
23. DESPATCH OF AREA MAPS TO THE URBAN AUTHORITY AND DELIMITATION OF - - URBAN AREA
24. PUBLICATION PROGRAMME -
25. COMPUTOR INSTALLATION AND PROGRAMMING AND TESTING - - - - - - - - -
26. DISTRIBUTION OF PORTFOLIOS, BALL-POINTS, PENS, HOUSING CENSUS MATERI- - ALS TO FIELD OFFICERS
27. TRAINING OF C.D.O. CHARGE SUPDT. CIRCLE SUPERVISORS AND ENUMERATORS- FOR HOUSING CENSUS
28. HOUSE NUMBERING HOUSING CENSUS -
29. RETURN OF HOUSING CENSUS MATERIAL FROM FIELD OFFICER TO DATA PRO- - CESSING CENTRE
30. RECRUITMENT AND TRAINING OF PUNCHING STAFF - - - - - - - - - - - - -
31. TRAINING OF C.D.O. CHARGE SUPDT. CIRCLE SUPERVISORS AND ENUMERATORS FOR POPULATION CENSUS
32. DISTRIBUTION OF BIG COUNT MATERIAL TO THE FIELD STAFF - - - - - - - - -
33. BIG COUNT -
34. RETURN OF FILLED IN CENSUS QUESTIONNAIRES FROM THE FIELD TO DATA - PROCESSING CENTRE
35. POST ENUMERATION QUALITY CHECK - - - - - - - - - - - - - - - - - -
36. SAMPLING -
37. EDITIND AND CODING -
38. TRANSCRIPTION OF DATA TO TAPES - - - - - - - - - - - - - - - - - - -
39. TABULATION (BY COMPUTOR) -
40. PRINTING AND PUBLICATION -

FIG 3.1 *Calendar for census procedures: Bangladesh, 1974*
(*Source*: L. J. Cho (ed.), *Introduction to Censuses of Asia and the Pacific, 1970–74*, Honolulu, 1976)

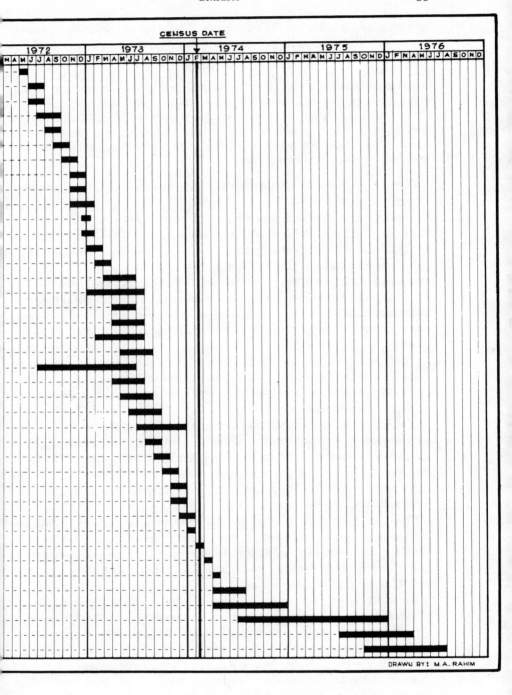

CENSUS DATE

1972	1973	1974	1975	1976
MAMJJASOND	JFMAMJJASOND	JFMAMJJASOND	JFMAMJJASOND	JFMAMJJASOND

DRAWN BY: M.A. RAHIM

footpaths in the vicinity, the supervisor should ensure that the enumerators agree on which one to use. In Sierra Leone, for example, it is reported that:

As a result of the urban nature of most parts of Western Area, it was decided to let E.A. boundaries run down the middle of the streets. The enumerators had no problems identifying these streets in the Western Area. In other parts of Sierra Leone . . . there were occasions where the E.A. boundaries could not be decided without becoming involved in chiefdom and village boundary disputes . . . the disputed area was noted and the E.A. boundary arbitrarily placed.[4]

Other aspects of enumeration are covered in later chapters, as they apply to censuses and surveys alike.

The concepts and definitions used will not be peculiar to the census, since they are likely to become precedents for later surveys. The following general, and rather obvious, principles, should, therefore, be observed:

(a) Concepts and definitions should be chosen for their relevance to national needs, and adapted to the local situation and tradition.

(b) Regional differences within the country need careful consideration. Language and cultural variations should be assessed and a conscious attempt made to ensure consistent and unambiguous responses across the country.

(c) Although experience and the study of work in other countries can be valuable guides, no one can foresee all the misunderstandings, etc., that can arise from the wording of the proposed questions. Pre-testing and post-census evaluation are necessary.

The people in the best position to evaluate a census are the statisticians who organized and analysed it. If no account of an evaluation is given, then some suspicion of the census must inevitably arise. One difficulty is that the publication of a full account of the evaluation does not usually occur until the end of the whole census operation, often years after the basic tabulations have been made available.

Once actual census results are available, checks can be made by examining external and internal consistency.[5] A wide repertory of such checks is available for population data, where the range of variation is relatively circumscribed by biological and social constraints, and the variables are closely interrelated.

Comparison of the census with other data will serve as a test of consistency: for example, figures of child population in the census may match returns from the educational system. If both sets of information agree, it does not necessarily mean that both are correct since they could be similarly in error. Nevertheless, if the two sets of information come from different organizations and/or were collected in different ways, then consistency will be regarded as helping to confirm them both. The two sets of information must be independent.

Extensive research resulting in a variety of checks on internal consistency has been carried out since 1950, and the work of Brass and Coale is particularly noteworthy.[6] Are the fertility data consistent with the age structure of the population? What is the extent and character of age heaping? Does the pattern of migration show compensating effects in urban–rural patterns by age and sex? There are also a number of well-attested common errors, such as the generally occurring under-estimation of very young children. In addition to straightforward data checks, a range of models is available to search for consistency, and to provide a set of estimates of demographic relationships that make the best use of the data available. These analytic methods depend on expectations about the type and order of demographic processes, which are derived from an accumulation of population data from many sources over a long period. Thus, the age structure found at a census can be checked against a range of model stationary populations, each of which can be generated by a number of fertility experiences. None of these methods compensates entirely for defects in the data; indeed their sophistication may sometimes induce too ready an acceptance of their results. They should in any case be supplemented by sensitivity analyses: that is, the rates or ratios emerging from them as the most likely should be given a range of possible variation, and estimates and projections recalculated with these upper and lower bounds to see how far data-based decisions would be affected by variations within the range. Many of these analytic techniques of estimation can be strengthened when the census to be evaluated is part of a series of censuses.

The most recent development is to consider the data needs of these techniques when determining the questions for the census. One proponent of this indirect approach writes:

Instead of asking 'What data are needed in order to calculate x, the variable under study?', a rather different question is asked, namely, 'What data, influenced among other things by the variable x, can we collect with reasonable accuracy?' The data thus collected will often not be pure measures of x, since other variables not of immediate interest may also affect them. Use is then made of theoretical or empirical demographic regularities in order to control for the influence of the extraneous variables, and thus derive a pure estimate of x. These regularities have very often been incorporated in the form of demographic models, whether of fertility, mortality, or nuptiality, and such models have played a central role in the development of indirect methodology.[7]

If consistency checks are combined with background knowledge about the country, it should be possible to reach some general conclusions about the strengths and weaknesses of a census. Since the census will serve as a frame for future activities and a bench-mark against which further data are monitored, this assessment is a continuing procedure: an obvious example is the way estimates of coverage deficiency at a census are usually revised when data become available from a subsequent one. We do not wish to provide an easy excuse for error, but it may be worth noting that in the 1940 census, after 150 years of census taking, the net census undercount of non-white men in the US was of the order of 20 per cent.[8]

3.3   THE CENSUS AS FRAME AND BASELINE

The characteristics of a sampling frame are discussed in Chapter 4. Here we note, in advance, that the census rapidly gets out of date as a frame for households, and gradually gets out of date with respect to dwellings and structures (particularly in the sprawl of poorer housing in urban areas). The value of the census for frame construction derives especially from the mapping work that is undertaken for it, and the related identification and clarification of boundaries of small areas. The census framework of enumeration areas is, or should be, very useful for later studies; its value is increased if the effectiveness of the demarcation of enumerator areas is assessed and recorded after the census enumeration. The words 'should be' unfortunately have to be added in the last sentence, because difficulties persist in the mapping and listing processes. Advances are being made using maps derived from satellite imagery and aerial photography, but regular updating of the listing of households within each small mapped area is a task beyond the capacity of most administrations. Censuses are launched in which the enumerator is responsible for locating each household and individual within an area that is mapped only to the extent that the boundaries are marked. Even these are often difficult to identify unambiguously on the ground. Using the census enumeration areas for selection of the samples in subsequent surveys requires a relisting of the households in the selected areas. There is no doubt that inadequate maps are a major source of error in censuses. Unfortunately, even expensive efforts in some recent censuses to improve the mapping have not achieved the virtual elimination of such errors. A post-enumeration sample check of enumeration areas following within weeks of the population census of Kenya in 1979 revealed major differences in household counts in individual areas due to boundary ambiguities. Yet a major mapping exercise had been undertaken for the census.

  If there is an explicit plan to introduce a national sampling scheme after the census, using the census as a frame, then a master area sample can be drawn and special care given to the documentation of the selected areas. The number of areas selected should be large enough to allow for rotation of areas over the years of the national sample plan. Only the location and the population for these areas may be available for delimiting strata, since the socio-economic data from the census will usually be unavailable until later. If the census operation is a combination of census and sample, and area sampling units are one stage in the sampling scheme, then it may be that the census sample areas can be specially covered, and used for the long-term national scheme. Clearly the possibilities depend on resources, but this kind of consideration reinforces the general arguments of this book for a long-range assessment of statistical resources and their co-ordination into a long-term programme of field investigations.

The field experience obtained during the census will be of value. Major difficulties in defining households will have been identified and should have been clarified in a coherent fashion. Difficult areas will have attracted attention. Experience in questionnaire design and of the qualities of enumerators should help in the design and costing of future survey work. The best enumerators may be attracted to full-time work, and there should be an increase in the pool of experienced workers available. In the first instance, the benefits of this extended experience may be limited to a few statisticians connected with the census. But the proper reporting of the census will make part of it available generally; and, provided good relations are developed between official statisticians and other research workers it will gradually percolate through to most people who are (or are likely to be) concerned with field inquiries. As with all attitudes that spread in this unregulated, rather haphazard way, some misleading 'folklore' or rules of thumb may accumulate, but this is a small price to pay.

The codes used for area and socio-economic variables (education, language, tribe, religion, etc.) will provide a basis for classifications used in future inquiries. The data will provide a basis for stratification and planning of further surveys. Cross-classification will provide control totals, and interrelationships and ratios will provide supplementary information and weights which can be used to provide more efficient estimates in future sampling plans. It will often be necessary to project the census figures forward for these purposes; and it must be remembered that the reliability of these projections deteriorates over time.

We have suggested that too much weight has been given to census taking, and that when censuses are taken they should be combined with sample inquiries. Nonetheless, we do not underrate the role played by the census, particularly a population census, in the development of statistics in a country and in its general modernization. An operation of this kind, involving the whole administration and people in a serious exercise highlighting numbers and accuracy, is an essential part of general social education and a contributory factor to nation building. This was sometimes recognized by national leaders even in colonial times. A Chandra Sekhar, Census Commissioner in India, records:

The earlier censuses of the country under the British did indeed raise various suspicions. . . . But the enlightened realised the great value of the census. Census taking in 1921 and 1931 faced serious difficulties due to the non-co-operation movement that had been launched. . . . It required the wisdom of the Father of the Nation, Mahatma

Gandhi, who, despite the freedom struggle he was leading, appealed to the nation to co-operate with the government in the taking of the census. . . . The census came to be recognised as a national undertaking of great importance and the persons called upon to perform the census tasks deemed it a national obligation to fulfil that responsibility.[9]

The benefits accruing from a successful census to the statistical office itself are substantial, and put it in a stronger position to ask for and expect resources and co-operation. We should perhaps repeat that, despite our emphasis on sampling, we consider that a periodic population census activity continues to be an essential item in any national statistical programme.

3.4   OTHER CENSUSES

Next to a programme of population censuses, the most commonly advocated is a census of agriculture. The FAO has sponsored the decennial World Census of Agriculture Programme since its inception. As originally conceived, it was intended to be a genuine census of holdings. Panse wrote in an FAO manual, 'straightforward agricultural statistics, such as area and yield of crops, number of livestock, sale of farm produce etc., may be regarded as being of secondary importance in the agriculture census.'[10]

Unless carried out in developed countries, which could use the system of mailed questionnaires to be completed and returned by every holder, the Census of Agriculture Programme was always a misnomer. Most countries claimed participation on the basis of a sample survey. Moreover, in most instances, the primary objective became the estimation of the current levels of cultivation, husbandry, and production. Gradually the FAO published guidelines which recognized this and gave greater prominence to what Panse had referred to as those statistics 'of secondary importance'. The FAO sum up the situation in their Programme for the 1980 World Census of Agriculture:

A few decades ago the agricultural census was considered the main source of all types of agricultural statistics. The increased need for more current data and the development of national agricultural and other statistical services and of the application of sampling techniques have resulted in a distinction between structural data to be collected through an infrequent periodic agricultural census and that to be collected more frequently through periodic and ad hoc statistical enquiries. . . . This requires a thorough examination of data requirements in relation to the whole system of food and agriculture statistics. . . .[11]

The Programme suggests that the population census and labour force surveys may be used to obtain basic information on the demographic characteristics of holders' households and on employment in agriculture.

It notes that sampling may be the most practicable method in many countries. One problem with a full census is that the minimum size limit set for holdings to be enumerated, dictated by the scale of the exercise, excludes a large number of the smaller holdings with which development programmes are concerned. We shall not explore the agriculture census further, therefore, assuming that most developing countries will select the option of sampling; and problems relating to agricultural sample surveys are dealt with in other chapters.

The type of census in developing countries already discussed involves enumeration by interview and observation as these are the only ways to obtain the information in the circumstances prevailing. There are, however, a number of sectors that can be covered by a mail questionnaire, with supplementation where necessary by interview. These sectors contain organizations that have the records and personnel from which they can provide the data required. Within the agricultural sector they are the larger farms, some of which may be owned by companies and run by farm managers. In the mining, industrial, and service sectors they comprise all but the very small units. The small units are often very numerous; but they usually account for only a relatively small proportion of total production. It will generally be necessary either to exclude such small units from the population to be studied, or to cover only a sample of them by special procedures.

As in any inquiry a list of units is required. The basis for such a frame may be obtained from lists of licences issued by government offices and local authorities, post office lists, advertisements, etc. A resulting collated list can be checked against rating returns, by a listing of dwellings in sample areas, and by the data collected for housing censuses. This work may be simplified by restricting inquiries to urban areas, but such a limitation results in very poor coverage of activities such as mining and saw milling. Two features of the commercial scene give particular trouble. One is that frequent small differences in names result in several names remaining on the list representing one unit. The other is that many of the units operate only intermittently (especially in building and trade) and there is doubt as to when names should be deleted. Both these features cause the frame to be inflated. This inflation is likely to be offset by units omitted because they have not been recorded in any of the sources. As the list is built up a regular investigation is needed to improve and update it.

The treatment of organizations can vary according to their size. Within any legal and administrative framework, the smaller units will generally keep less full and accurate records than the larger. Less information needs to be obtained from them. Further, a full census need be taken of only the larger firms since the smaller units can be adequately covered by a sample. Such a combination of census and sample can include different methods of data collection, using interviews for smaller firms either from the outset or as a first follow-up procedure. The dividing line for the strata can be the same for all activities or vary from one activity to another. Thus, although a lawyer's operations may be small, he may well be able to provide information about his activities in a way a shopkeeper of the same importance could not match. The extent of small-scale activity in distribution makes it a particularly difficult sector to deal with.

One of the most demanding examples of censuses of businesses is provided by the annual censuses of companies and industries in Malaysia.[12] These censuses have considerable political support for they provide the basis for monitoring progress under the New Economic Policy of the government. The censuses cover over one hundred thousand businesses, many of which need to be visited in order to provide assistance with the completion of the form. This example illustrates the negative aspect of such annual censuses: it places such demands on the statistical services that other priority surveys are bound to be adversely affected.

The items of information that a statistician can hope to collect depend upon the tax system and the accounting conventions. It is often difficult to establish what the financial conventions are (for example, in the treatment of investment), and whether they are consistently applied throughout industry. The mining and building industries and the distributive trades, in particular, give great difficulty. Discussions with practising accountants, trade associations, and officers in the tax department are necessary. Economists and statisticians involved in national accounting also have to be consulted, since one of the main purposes in collecting the information is to provide a sound basis for the national accounts. The national accounting framework with its built-in structure of cross-checks is one of the few ways of testing the data for consistency.

In addition to these unavoidable difficulties with concepts, a major obstacle to overcome is the suspicion of the companies that any information they give may be made available to the tax authorities with subsequent disadvantageous results (to the company). It is very difficult to remove this suspicion. Whilst doing all that can be done, especially through trade associations and chambers of commerce, to convince operators that confidentiality will be maintained, it may be necessary in the first instance to avoid direct questions on the most sensitive areas, such as profits. Suspicions are often more acute with the smaller units which may not have regularly audited accounts. These small operators may include many of those potential entrepreneurs which the development plan and the ministry dealing with commerce and industry are trying to encourage.

Information from 'commercial' agricultural organizations will have to be co-ordinated with that collected directly from small farmers. Studies relating agricultural output to its processing in industry and its eventual marketing at home or as exports will also be needed. Generally, however, the work discussed in this section requires a somewhat different background and approach from that necessary for the work discussed in the rest of this chapter (and much of the book). Within the framework of a national statistical organization, senior statisticians will need experience of both types, although it is not advisable to move personnel around too much early in their careers. There is one danger. Censuses of business organizations will be mainly in urban areas, and will provide those engaged in them frequent opportunities for involvement in economic and financial planning. Consequently, this work may appear more pleasant and influential; and the heavy field involvement in household and holding investigations, often in remote parts of the country, with poor transport facilities, may make these assignments less popular. Steps should be taken to adjust rewards so that this imbalance does not result in the field enquiries losing—or never receiving—good staff.

Notes

1. El-Khodary, M. A., 'Problems of Demographic Data Collection in Arab Countries', *Population Bulletin of the UNEC for Western Asia*, vols. x–xi, 1976, pp. 91, 93.
2. Al-Adashi, A. K., 'Some Technical Problems Encountered in the Population Census in the Yemen Arab Republic', *Population Bulletin of the UNEC for Western Asia*, 1976, vols x–xi, p. 182.
3. Suharto, S., McNicoll, G., and Cho, L. J., 'Indonesia', in Cho, L. J. (ed.), *Introduction to Censuses of Asia and the Pacific, 1970–74*, East–West Population Institute, Honolulu, 1976, p. 79.
4. Okoye, C. S., *Population Census Interviews*, Fourah Bay College, Freetown, 1976, p. 5.
5. UN, *Handbook of Population and Housing Census Methods*, UN, New York, 1969–72, 7 parts.
6. Many of these methods have originated from and been synthesized by W. Brass and A. J. Coale. See Brass, W., *Methods for Estimating Fertility and Mortality from Limited and Defective Data*, International Program of Laboratories for Population Statistics, Univ. of North Carolina Press, 1975; and UN Manual IV: *Methods of Estimating Basic Demographic Measures from Incomplete Data*, Population Studies, No. 42, UN, New York, 1967.
7. Hill, K., 'Census Data Required for Indirect Methods of Estimating Demographic Parameters—1980 Round of Censuses', *Population Bulletin of the UNEC for Western Asia*, 1977, vol. xiii, p. 57.
8. Coale, A. J., and Zelnik, M., *New Estimates of Fertility and Population in the United States*, Princeton Univ. Press, Princeton, New Jersey, 1963.
9. Chandra, Sekhar, A., 'Population Censuses and Vital Statistics in India', *International Population Conference, Liege, 1973*. IUSSP, 1974, vol. ii, p. 344.
10. Panse, V. G., *Some Problems of Agricultural Census Taking with Special Reference to Developing Countries*, FAP, Rome, 1966, p. 7.
11. FAO, *Programme for the 1980 World Census of Agriculture*.
12. Annual Reports of the Department of Statistics, Government of Malaysia.

4

Sample Surveys

Observe the small facts upon which large inferences depend.

The Sign of the Four

4.1 THE ADAPTABLE TECHNIQUE

The sample survey provides a flexible method that can be adapted to almost every requirement of data collection. The case study does not provide for direct inference to a population; the census does not require an inference because by definition it covers the entire population. The sample survey covers the many circumstances in which inferences about populations are required, but a census is either not possible or desirable.

The advantages of sample surveys can be summarized under three headings:

(a) Economy: this includes economy of cost and of time, because only a limited number of units have to be examined and analysed.

(b) Accuracy: because the quality of enumeration and supervision can be higher than in censuses, the quality of the data collected should be better. This improvement will generally more than offset the variability in the results arising from the sampling process.

(c) Adaptability: many topics, particularly those involving detailed transactions of individuals or households require an intensity of interview or observation that cannot conceivably be covered by censuses. A sample is the only mode of inquiry available.

One of the results of the sampling theory, developed in the manner outlined in Chapter 1, is that samples from which inferences are to be drawn have to be selected so that each member of the population being studied has a known, non-zero probability of inclusion in the sample. This will involve randomization at one stage or another. Some of the problems of selecting a sample are dealt with briefly in this chapter. But first we develop and apply a typology of surveys suitable for considering the main characteristics and organizational background of survey work in developing countries.

4.2 A TYPOLOGY OF SAMPLE SURVEYS

Sample surveys have been classified by various criteria. One dichotomy often regarded as fundamental is that between the descriptive and the analytical survey. 'Descriptive' describes itself; Cochran defines an 'analytical' survey as one where 'comparisons are made between different subgroups of the population, in order to discover whether differences exist among them and to form or to verify hypotheses about the reasons for these differences'.[1] He comments that the distinction between the two types of survey is not clear-cut and that many surveys serve both purposes. This dual function is especially common in surveys in developing countries, and we think it more appropriate in the present context to use categories relating to the scale of the survey and the intensity of the collection process. The following classification lists five criteria, each criterion being scaled by approximate divisions.

Criterion	*Classification*		
	(a)	(b)	(c)
1. Geographical area of coverage	Village(s)	District/Region	National
2. Number of respondents	Less than 100	100–1,000	More than 1,000
3. Subjects of study	Single	Multiple (integrated)	Multiple (successive)
4. Frequency of enumeration	Single visit	Multiple visit	Regular, frequent, continuous
5. Data collection procedures	Interview	Interview and measurement	Observation

The dividing lines between entries in columns (a), (b), and (c) are rather arbitrary divisions of continuums. Clearly, there are not sharp breaks at the specific numbers 100 and 1,000 in the range of number of respondents (Criterion 2). The division between cells 4(b) and 4(c) depends on how frequent is 'frequent'. A district in one country may be larger than a neighbouring nation. The way in which we intend these dividing lines to operate will become clear, we hope, from the following paragraphs.

In Criterion 3 we use 'single subject' and 'multiple subject' to describe surveys involving the investigation of one or several subjects respectively. The terms 'single purpose' and 'multi-purpose' are sometimes used, for example in a recent publication of the Organization for Economic Co-operation and Development.[2] The UN definition, quoted by Murthy, states that a 'multi-subject survey is a simultaneous investigation of several subjects not necessarily closely related in a single survey operation for the sake of economy and convenience . . . data need not necessarily be obtained for the same set of sampling units or even the same type of units'.[3] This definition covers a very wide range of possible activities, some of which would be extremely difficult to operate. Thus, if the data for the different subjects are to be collected within the same survey round or phase we strongly recommend that the subjects *should be* at least indirectly related and the procedures to be used for collecting the data should be common. To combine a subject requiring a recall period of two days with one that needs only a monthly visit is to invite a compromise procedure that is less than satisfactory for both. Any interview, if interview is the adopted method, should follow a logical course and not involve too abrupt a switch of topics. Subjects that are completely unrelated cannot be handled appropriately within a single questionnaire. The true multiple-subject survey, to our minds, is one in which data relating to several related subjects are collected from the same respondents using an integrated questionnaire and the same basic framework. This is referred to in the above typology as multiple subject (integrated). Different subjects may also be studied in different rounds of a regular, frequent, 'continuous' survey. Here there is no requirement that the respondents are the same, but it is assumed that successive rounds are carried out within the framework of the same national sample. We refer to such surveys as multiple subject (successive). If each subject is studied on a different sample of respondents within the same reference period, we consider the operation as a set of single-subject surveys.

The relative merits of carrying out surveys covering a range of distinctly unrelated subjects in single or successive rounds remains a matter of controversy. In recent years the UN-sponsored Household Survey Capability Programme has assisted many countries in developing a programme that fits our definition of multiple subject (successive) surveys, i.e. annual rounds within a master clustered sample that cover individual subjects in successive years. But some countries, assisted by the World Bank, have experimented with an integrated approach to living standards measurement in which many subjects are covered in one round of a survey—these subjects being as varied as fertility, employment, income, and nutritional status.[4]

Surveys requiring only one visit to each respondent are highly desirable, allowing a large size of sample for a given cost. Many surveys, however, require a series of visits if the objectives are to be met. Such multi-visit surveys are the norm rather than the exception for a wide range of agro-economic surveys. For some purposes, visits may have to be so regular and frequent that the data may be regarded as being collected on a continuous basis.

The required information may be collected by interview, by measurement, or by a combination of these procedures. The third possibility is to observe the respondent or the phenomenon, recording the results of the observation. Such a procedure is sometimes used in a continuous data collection survey.

The typology given above provides 243 cells, but many of them are 'empty', that is, they are of no practical significance. It is unlikely that a single village sample survey would include more than 1,000 respondents or that a national survey would have fewer than 100. Thus, cells deriving from the combination of $1(a) \times 2(c)$, and of $1(c) \times 2(a)$ are of no importance. The usefulness of a typology combining the listed criteria is that most surveys fit into one of a very few of these cells, although there will clearly be exceptions. A few examples are shown below.

A. *The case study* (see Chapter 5)

Type $1(a) \times 2(a) \times 3(a)$
This is suitable for an individual research worker or for a research team, but, except in pilot inquiries, will not normally be carried out by official data collecting agencies. The simplest investigation of all within this type is $1(a) \times 2(a) \times 3(a) \times 4(a) \times 5(a)$, involving a single interview on one subject; this is often used to provide quick information on a very narrow front or for a pilot study to test a questionnaire.

A more common type of case study, for example a farm management study, is $1(a) \times 2(a) \times 3(a) \times 4(b)$. Several visits are paid to each respondent, and the procedure involves a mixture of interview and measurement $(5(b))$.

The most specialized case study involves continuous observation of the units of study; that is, a survey of type $1(a) \times 2(a) \times 3(a) \times 4(c) \times 5(c)$: this requires a highly trained team with professional knowledge of the topic under study, for example, a dietary survey.

B. *District or regional* ad hoc *sample survey*

The simplest survey of this type, $1(b) \times 2(b) \times 3(a) \times 4(a) \times 5(a)$, is similar to the case study; but when based on a probability sample allows inferences about the parent population of a district or region to be drawn.

C. *Evaluation survey* (see Chapter 14)

Type $1(b) \times 2(b) \times 3(a) \times 4(a)$
Its use for a *post-hoc* evaluation of a specific development project will normally require an evaluation team's using a combination of interview and measurement $(5(b))$.
Type $1(b) \times 2(b) \times 3(b) \times 4(b)$
This is suitable for monitoring and evaluating an ongoing integrated rural development project and assessing its social and economic impact.

D. *Urban or district budget survey* (see Chapter 11)

Type $1(b) \times 2(b) \times 3(a) \times 4(b) \times 5(a)$
This is an example of the standard limited area sample survey conducted by official data agencies and also by institutions.

E. *Experimental trial survey*

Type $1(b) \times 2(b) \times 3(a) \times 4(c) \times 5(c)$
This is an inquiry similar to those conducted in a clinic or research station,

but using a probability sample of a population: for example, fertilizer trials, child growth–diet survey.

F. *Survey of a subset of a population*

Type 2(b) × 3(a) × 4(b)
This may be conducted by researchers into topics applicable to subsets of a population: for example, a survey of market traders, or a study of rehabilitated patients. The geographical coverage may be regional or national (1(b) or 1(c)), but the sample will not normally exceed two hundred or so.

G. *Standard small-scale, multiple-subject sample survey*

Types 1(b) × 2(b) × 3(b) × 4(a) × 5(a)
     1(b) × 2(b) × 3(b) × 4(b) × 5(a) (or 5(b))
On a small scale, covering a district or region, such a survey is still within the scope that can be contemplated by a research institute; but the multi-subject survey is generally better suited to an official data collection agency utilizing an experienced field force.

The first option given above risks overloading the single interview. Use of this option certainly requires that the subjects are related: for example, a survey dealing with fertility and attitudes to contraception may be viewed as a two-subject survey on closely related topics.

Many multiple-subject surveys require multiple visits. A budget and food consumption survey is an appropriate example; either interview or measurement techniques may be used.

Type 1(b) × 2(c)
In countries that are large, both geographically and in terms of population, even a subnational survey may include more than 1,000 respondents.

H. *National* ad hoc *sample survey*

Type 1(c) × 2(c) × 3(a) × 4(a)
National level surveys are usually conducted by the official data agencies. Type 1(c) × 2(c) × 3(a) × 4(b), that is multiple visits on a national scale on an *ad hoc* basis, is expensive, and consequently difficult to justify for this reason.

I. *National integrated sample survey*

Type 1(c) × 2(c) × 3(b) × 4(a) × 5(a)
     1(c) × 2(c) × 3(b) × 4(b) × 5(a) (or 5(b))
This is a multi-subject survey similar to G, but on a national scale. The size and spread of the sample and the length of questionnaire normally required are likely to lead to low data quality if the single visit option (4(a)) is used.

Multiple visits with measurements and interviews require an extremely efficient organization.

J. *National continuous sample survey*

Type 1(c) × 2(c) × 3(c) × 4(b) × 5(b)
A permanent survey capability operating a continuous programme of surveys in which single-subject surveys are conducted successively at intervals.

K. *The census* (see Chapter 3)

Type 1(c) × 2(c) × 3(a) × 4(a) × 5(a)
The census (in the sense of a 100 per cent sample) is the extreme case of the 1(c) × 2(c) type. Single subject 3(a) is recommended, interpreting 'single subject' to include the basic particulars of individuals, including education and occupation. The inclusion of variables such as income would put the census into type 1(c) × 2(c) × 3(b) × 4(a) × 5(a); we do not recommend this in developing countries.

L. *Special examples*

Type 1(c) × 2(c) × 3(a) × 4(a) × 5(a)
A special case of this type is a mail questionnaire. The single visit, single interview is not by an enumerator but by the surveyor in the form of written questions.
Type 1(b) × 2(b) × 3(a) × 4(c) × 5(c)
The continuous observations are recorded by the respondents themselves, that is, a diary-recording survey where the diary or account book is left with the respondents, to be completed by them.

The above examples cover the most common types of surveys used and recommended for the data needs of developing countries. We do not suggest of course that surveys lying outside one of our selected types must be 'bad' surveys. We would recommend, however, that any surveyor who finds his survey fits into one of the other cells of the typology, should take a second look at his plans, to see whether one of the more common types will not be more appropriate.

4.3 *AD HOC* OR PERMANENT SURVEY PROGRAMMES

An *ad hoc* survey is conducted as an entity in itself, aimed at meeting specific objectives within a certain time span.
 Characteristics include:
(a) The sample design and selection procedures are chosen to meet the specific needs of the single inquiry.

(b) The enumerator and other survey staff can be chosen from those possessing the appropriate skills relevant to the inquiry.

(c) The timing of the survey can be that most convenient for obtaining the data.

(d) The entire operation is under the control of a single surveyor or very small team.

Although, as discussed later, some new surveys with limited and precise objectives can ride 'piggyback' on a continuing sample programme, many surveys require careful choice of an appropriate design and selection procedure if the data collected are to meet the objectives. This will be true particularly when the phenomenon being studied is uncommon, or the population of interest to the surveyor is a particular subset of the entire population and is not randomly distributed within it. A survey to assess the incidence of a rare disease, or a survey of itinerant food traders, cannot be conducted using a sample selected for broader and more general purposes. An *ad hoc* survey allows the surveyor to choose his staff according to the topic, and he can usually choose his timing. A survey of school leavers must be timed around the end of the scholastic year. An opinion poll of the impact of budget changes should be conducted shortly after they are known. A survey of traders dealing in rural markets must be timed according to both the day of market operations and the season of the year.

Many private researchers are anxious to have the entire survey under their own personal control. They consider, usually correctly, that no one else views their concerns with the same intensity; a sample and enumerators controlled by another person or agency leaves them exposed to delays or complications or indifference that they can do little to avoid or influence.

However, the *ad hoc* survey can suffer from the following disadvantages:

(a) The cost per respondent enumerated is high.

(b) A suitable frame may need to be constructed.

(c) The enumerators may lack experience and motivation.

(d) Professional and administrative support is difficult to obtain.

Setting up an individual survey is time consuming. The occasional surveyor is faced with the task each time he embarks on an *ad hoc* survey. Inevitably costs will be high relative to the cost per topic in a multi-subject survey operation, and may result in the skimping of necessary preparatory work and the cutting of infrastructural support below an acceptable level. The inevitable result is a deterioration in data quality.

The construction of a frame from which to draw the 'tailor-made' sample may take considerable effort. The effort may be well spent, for this process in itself may give the surveyor new insight into the topic he is investigating. It may be possible to construct a frame of urban retailers only after a lengthy process of searching and classifying the records of the business licensing office, supplemented by inquiries of the local council and a summary of the returns submitted by all retailers to the ministry controlling trade. Even then the frame may still be inaccurate and incomplete, making substantial verification on the ground necessary. The results of these inquiries and verification will clarify the surveyor's mind, and enable him to establish his objectives more precisely by showing him the area in which he can hope to work successfully.

Novice enumerators require extensive training and practical experience before the quality of data collected is fully acceptable. The experience gained by enumerators in one *ad hoc* survey may be lost to the surveyor forever, since these 'veterans' may be unavailable at the time of the next survey. If the number of respondents is small enough to be enumerated by the surveyor and one or two permanent assistants this disadvantage does not arise, but the size of sample will be, by definition, severely limited.

The surveyor may need help in the statistical design of his survey, the drafting of the questionnaire, the collection of the data, and the preparations for data processing and analysis. The occasional surveyor may encounter difficulty in obtaining such skills and field support, although these problems can sometimes be overcome by building up contacts with the relevant agencies.

Many of the disadvantages of the *ad hoc* survey disappear if the total survey programme is sufficient to create and maintain a permanent survey organization. In the early days of statistical development surveys are carried out sporadically; periods when surveys are in operation are followed by periods when no such activity is taking place. At some point, however, the needs and resources become large enough to justify a permanent survey force. Even at this stage problems remain. Surveys tend to be financed individually by various ministries and technical departments, and the survey operations may be decentralized with each ministry or department working to meet its own particular needs. But over-all national priorities should eventually prevail. A permanent field force of supervisors, enumerators, and data processors will allow for the accumulation of experience, and offer the opportunity for regular employment with some prospect of advancement that will improve morale.

The creation of such a permanent survey capability should improve efficiency. Although the range of surveys may be wide, the existence of a corps of experienced surveyors with a stable field environment provides an opportunity for standardization of concepts and definitions. The user who needs to review the results of various surveys is often hampered or misled by the different concepts and definitions used in individual surveys. Data from two surveys that in theory refer to the same population may be found, on examination, to be incompatible owing to different definitions of the sample unit. Two units of key importance, the household and the holding, are particularly subject to varying definitions and are discussed in Chapters 11, 12, and 13.

The work of the permanent survey team should also lead to the introduction of standard procedures, not only within the survey organization but also, because of the force of example and the benefits to be derived from comparability, in other survey work in the country. The techniques to be used for measurements of land areas and crop yields, the probes to be used for determining income, and the recall period to be used for different inquiries are all procedures for which standards appropriate to local conditions should be set.

The objective within a developing country, as we see it, would be for technical advice to be available to all surveyors from the pool of professional survey designers within one major survey organization, and for the field force of this organization to be used for as many of the large-scale general surveys as possible. The small surveys and those requiring a specialized type of enumerator, as discussed above, would still be handled by other investigators.

Once a permanent survey force exists and is being efficiently maintained the next logical stage of development is likely to be the creation of a national sample with a programme combining a basic 'core' survey with other surveys carried out concurrently or sequentially within it.

The first viable, well-documented national sample of this type in the developing world was achieved in India (see Chapter 1). The architect of the Indian National Sample Survey, P. C. Mahalonobis, stressed the economies in time and cost, and the improved appraisal and control of sampling and non-sampling errors made possible through the creation of such an organization.[5]

The first and most obvious feature of a national sample survey is that it is based on a sample structured and designed to give estimates within certain confidence limits at an agreed level of geographical aggregation for a wide range of variables, and not just for a few characteristics. Whatever type of sample design is used, the basic structure will remain unchanged for some years, during which time many surveys will operate within its basic framework.

The type of sample commonly adopted for a continuous national survey is a stratified, multi-stage clustered sample. The primary sample unit is usually a small administrative area. The frame will almost certainly be related to the areas demarcated in the population census, but care should be taken not to choose too small an area in view of the need to rotate respondents within the sample area clusters. Once the selected primary units have been identified and mapped, further stages of area sampling can be incorporated, leading to the selection of a conveniently sized cluster within the primary unit. The population of final sample units in this cluster are listed and a sample chosen from the list. Various stratifications can be used at one or more of the sampling stages. The stratification must use general characteristics as criteria, since the national sample will serve many purposes; stratifications related to special characteristics may not be of much help generally. The selection of the sample units at each stage can be made in various ways, but must meet the essential condition of randomness if estimates for the total population are to be calculated from the sample data.

Clearly the initial sample design is of vital importance. Undue complexity, or optimistic assumptions about the accuracy of frames or maps or initial listings may result in biases that will affect all the surveys using the national sample. Each extra stage of sampling, each refinement of stratification, or weighting of probabilities of selection, increases the risk that something will go seriously wrong in the execution of the design on the ground.

The most attractive feature of the continuously maintained national sample survey is that data for many purposes can be collected in a complementary manner that cannot be achieved by a series of *ad hoc* surveys. The major topics that need to be surveyed regularly can be integrated into a series of rounds in an appropriate order. These benefits accrue from the common sample structure, not from a fixed set of respondents—indeed, the respondents can be changed according to a fixed timetable. It is tempting to keep the respondents unchanged, but the imposition on them must be borne in mind, and experience shows that constantly surveyed respondents do not remain typical of the population that they were selected to represent.

But there are disadvantages as well as advantages to the use of national master samples in this way. Many countries participating in the UN Household Survey Capability Programme have chosen this option, but the tendency to equate a survey capability with the operation of successive surveys within a common master sample has been questioned. Clustered samples may be very inefficient for many surveys, particularly those related to small-scale agriculture. Frequent relisting of units within the sample areas is a requirement if the frame is not to become outdated. But, given this requirement, the option of drawing a new sample of clusters needs to be considered. Recent experience would indicate that the adoption of a fixed master sample has led to some surveys being forced to fit within the framework when an independent sample design was indicated.

Although the case for developing a national sample survey with a permanent survey organization to operate it is strong, instances of such a capability's growing and becoming more efficient over long periods of time are difficult to find. The efficient maintenance of such an operation requires a consistent commitment by the senior administrators and statisticians: this may be difficult to sustain once the initial impetus has been spent and the enthusiasm generated by a series of timely publications has given way to routine acceptance and expectation. Is the national sample survey a monster that finally destroys its makers? We believe the answer is no. Establishing a national sample survey brings problems in its wake—problems that cannot arise when that step has not been taken; but present experience suggests that working towards a national programme is still the best option available.

Moreover, it is important to understand that the problems stem largely from failure to meet infrastructural requirements. A national sample survey requires a permanent field force of enumerators and supervisors; regional offices adequately staffed to control this field force; a well-maintained supply of equipment and transport; good communications between the head office and the field; an efficient clerical staff to edit, code, and validate the data; a sophisticated data storage, retrieval, and processing system; a staff of statisticians and surveyors to oversee and control this operation on a continuous basis; and the capability to digest, analyse, and publish the information speedily, accurately, and completely.

It is unfortunately true that the necessary long-term support for these requirements is often lacking in the relevant planning and financial echelons of government. Survey capability is often an early casualty in times of budgetary cut-backs. As we shall argue later, some of the blame for this falls on the heads of statistical organizations: there has been a reluctance to focus survey work on the identified short-term needs of planners and those responsible for sectored development programmes. Even where such a focus has been achieved, deadlines have been missed. Data that are delivered after the decisions that required them are made have little value, and will receive little support for their collection in the future.

4.4   THE SAMPLE FRAME

The frame for a sample is a list of the units in the population (or universe) from which the units that will be enumerated in the sample are selected. It may be an actual list, a set of index cards, a map, or data stored in a computer. 'The frame is a set of physical materials (census statistics, maps, lists, directories, records) that enables us to take hold of the universe piece by piece.' Examples are a list of administrative areas, a file of industrial licences issued, a list of agricultural holdings or of households or of individuals in a village, a street map of a town. If we are choosing the first stage in a multi-stage sample the lists will not be of the final units examined and enumerated. Thus a common procedure is to draw a sample from a frame of villages, and then take a second stage sample of households in the selected villages. The frame of households has to be prepared only for the sample villages selected at the first stage. This is an example of 'area sampling' which has been defined as the method where 'the entire area in which the population is located is subdivided into smaller areas, and each elementary unit is associated with one and only one such area'.[7]

It may be necessary to distinguish between a target population of a survey, the population those commissioning or initiating a survey would really like to cover, and the population which is actually covered by the frame used in the final survey. This gap between the ideal and actual often arises either because no frame exists or cannot be economically prepared for the target population. For example, a survey to estimate the production of maize may have to neglect scattered production on plots in urban and peri-urban areas. In such a case, the gap may be of little significance; but it is not always a simple matter to decide how badly the gap will affect the results. For example, any survey of market gardening that neglected urban and peri-urban activity might well be so deficient that no useful inferences could be drawn from its results. The responsibility of deciding whether to go ahead with the possible rather than the desired population lies with those responsible for the decision to undertake the survey.

There is another facet to the issue of the population to be covered. Sometimes, it is not required that the survey cover the whole population in a given area. Those commissioning the survey may be concerned only with a part of the population, for example, those participating in a particular development activity or project (see Chapter 14). 'Formal sampling requires that a universe be identified, but the sampler is free to choose the universe . . . and it does not have to be all-inclusive.'[8]

Suppose we are confronted by a frame that has just been prepared, say a list of inhabitants of a village. What errors may it contain? It may be 'inaccurate' in the very general sense of that word. In discussing frames, however, it is common to break general inaccuracy or wrongness into three components: inaccuracy, incompleteness, and duplication.

Inaccuracy in a frame may be the result of a simple recording error, as when a name is misspelt, or a house is given an incorrect number, or an adult is recorded as a child. Or the recording may have been accurate but the information provided was wrong, because, for example, a neighbour provided information about a family absent when the list was prepared. Other errors may cause the list to be inflated; for example, households may be reported to be larger than they are. This kind of error can arise from deliberate or accidental misdirection, or from a misunderstanding about who is to be regarded as resident.

Such misunderstandings can of course work the other way, and lead to the exclusion of someone who should be included. A frame is then *incomplete*. Incompleteness can also arise from failure to include all the structures in the village when the frame is being created. Such a failure can arise from laziness on the part of the enumerator, or that he misreads the map, or does not check the village boundaries on his map against local knowledge. Thus Kannisto, reporting an underestimation of rural population in Cambodia, writes:

This was no doubt mainly due to lack of precision in the frame. Many hamlets, considered to be part of a main village, were not listed separately; the interviewer, however, hearing that a nearby locality had a different name, assumed it not to belong to the village assigned to him—and there were no detailed maps.[9]

In some cases the difficulty may be that the boundaries are not well established. For example, Hill, writing about Nigeria, records:

Hamlet boundaries are usually indeterminate where they cross uncultivated bush land and not necessarily determinate in well-cultivated areas: however, it must be added that the notion of indeterminacy is usually thoroughly repugnant to the parties concerned, each of whom claims to have his own view as to exactly where the boundaries run.[10]

and:

the boundaries of these Areas (especially the Hamlet Areas) are somewhat indeterminate: as one farmer put it, it is the general directions of the boundaries, rather than their actual positions on the farmland, which are known. . . . It is indicative of the lack of interest in Village boundaries that the most up to date official map . . . should fail to record the existence of Ciranci Village although it had been carved out of Ja'en some forty years ago.[11]

There have been many examples of surveys which have been badly damaged by failure to ensure that the boundaries of the units in the frame were precisely known and identifiable. We made a mistake in the sample census associated with the 1959 Uganda population census when we decided to use sub-parishes rather than parishes in the Buganda sample. Our preliminary investigations in the field suggested the sub-parish boundaries were known; but these investigations were clearly not on a sufficient scale since, in the event, 'the basic premise that the sub-parish areas could be rigidly controlled was not realized in practice'.[12] This boundary problem is less likely to arise in countries that have been well surveyed; but even then there is a danger that selection will move down one notch into unmapped subdivisions, as the statistician at the centre trying for an optimum design searches for small areas, or for some way to delimit areas of about the same size. By its nature incompleteness is difficult to uncover.

Duplication in the frame arises when the same unit is included twice. This can arise when the list is prepared by more than one person whose work overlaps, or when there are contested claims to land and more than one person lay claim to the same piece in their answers to questions. Duplication can sometimes be discovered by inspection; but work on matching connected with dual enumeration based on the Chandrasek-har–Deming technique shows how different spellings, initials, etc. lead to many non-matches and mismatches.[13]

Frames can be subject to these errors at the moment they are made. Even if they were entirely correct and complete at the time of making, they are soon rendered out of date as birth, death, and movement make them inaccurate and incomplete. The World Fertility Survey comments, for example, that 'in many countries expectant women habitually move to their mother's home to have their baby and return again to their husband's household a few weeks after the birth; thus if women are listed and then visited a month or two later for interview, we will selectively miss those who are in the last stages of pregnancy as well as those who have a very young child (the latter group were listed at their mother's home but have left it again by the time of the interview'.[14] Even a list of administrative areas in a country with stable government gets out of date as areas are subdivided or amalgamated or given changed boundaries.

Nation-wide frames are very seldom entirely prepared by surveyors, except in the context of major census operations; and even then they work within the framework of administrative boundaries and cartographic support. Many other frames have at least to start from records derived during some administrative process, such as taxpayer lists, land registers, etc. Any assessment of records of this kind as potential frames must take into account the system the record reflects, and the importance of the record to the administering agency. If there are disadvantages in being 'on the record', then the record is likely to be incomplete. Thus there is likely to be some, probably illegal, subletting of officially subsidized housing, so any list of households on supported estates is likely to be deficient. Indeed, there may sometimes be unofficial building on housing estates as well as unofficial renting. On the other hand, if being on the list conveys certain rights, for example, title to free or subsidized fertilizers, or general political advantage relating to numbers on the list, then there is likely to be inaccuracy (inclusion of persons who do not exist), or duplication of households or persons. One Caribbean finance department found it was paying fertilizer subsidies for an area greater than the total area of the island. The investigator has to obtain some general view of the situation in order to decide whether the record is likely to provide a useful basis for a frame, and to assess what errors have to be particularly searched for. Thus Stevens writes about an index of rural property owners which was being considered as a frame of coffee growers in São Paulo, Brazil:

With regard to the possibility of omissions, the a priori argument, if unscientific, is very strong. Registration of the property and regular payment of taxes is, to some extent, evidence of title. It is inconceivable that the owner of a valuable farm will try to keep it out of the register to avoid a relatively small tax which, in any event, will come to roost with compound interest added, when the farm is sold or transmitted to an heir. . . . On the contrary one worries more about duplication. It is known that the same piece of land is sometimes registered by two or more claimants.[15]

Because of the rapidity with which frames of households or individuals become out of date it is generally recommended that lists more than a few months old should not be used for selecting a sample of them. In increasing order of durability we move up through dwellings and small areas, to larger areas. This is one reason for the importance of the mapping and identification of enumeration areas in the census. This is likely to provide one of the safest and most durable frames of wide applicability, and it will be of reasonably small areas. There will be problems of updating, sooner or later, especially in peri-urban areas. The census data associated with enumeration areas will also be of great value in sample design.

One can check coverage by supplementary inquiries. For example, a list of firms can be checked by selecting sample areas or blocks, listing units operating in them, and checking back to see if they are in the register. Industrial activities are often zoned, and a quick tour of the town may be sufficient to confirm whether the zoning arrangements are being followed, so that the sample check can be improved by being concentrated in appropriate sections. A similar situation can arise in the selection of workers for a household budget survey. Such a sample is often obtained by selecting a sample of firms from a list of employers kept by urban authorities, the labour department, or the factory inspectorate. Such lists are likely to miss small employers, and it is dangerous to assume that workers in these small units have incomes and expenditures similar to those of workers in larger companies. It is therefore desirable to sample areas to check the adequacy of the list of operating units, and to take a supplementary sample of workers, whose results would be tabulated separately, from units which are found to exist but which are not on the original list.

A method which may be used when the frame has units serially numbered or sequenced in some physical order relies on the 'half-open interval'. This can be applied to dwellings in a street. As each sample number is chosen, a record is made of the next number appearing on the list. The enumerator is instructed to cover the sample unit and any dwellings existing up to that next number (account being taken of odd and even numbers according to the system of numbering in use). Stevens describes a similar check for farm properties as follows: 'An objective check on omissions is provided by listing the owners of all properties contiguous with all or some of the sample properties and subsequently trying to identify these in the register.'[16]

The existing frame may include a number of entries whose response to the intended inquiry may be 'nil'. For example, a list of farmers which is to be used to estimate coffee production may include a number who do not grow coffee. If some of these occur in the sample, their contribution to coffee production will be zeroes. These will (appropriately of course) make no contribution to the sample total, but will increase the variance of the estimate. With a multi-stage sample this can be avoided by a detailed listing within selected clusters. It should be noted that no substitutions should be made for sample units that give a nil response. This point should be emphasized in the training of supervisors and enumerators, who will generally be given no powers to make sample changes. The importance of recording zero when that is the proper response should also be stressed; and the field staff must be taught to appreciate that the situation is part of the over-all sample operation and that in the circumstances their time spent dealing with these cases is not wasted. It is also important to make sure that true zeroes are recorded and distinguished from 'not stated' or non-response situations.

Duplication in the sample may be treated by altering the selection probabilities allocated to the duplicate units, or—since altering probabilities may introduce difficulties at the analysis stage—by discarding at random one of the duplicates. It should be noted that the proportion of duplicates in the sample is lower than it is in the population. If f is the sampling fraction, then the proportion of duplicates in the sample is f times the proportion in the population.[17]

4.5 SAMPLE DESIGN

This section is particularly brief for our focus is on practical survey problems, which tend to be neglected. We do not wish that our particular emphasis should be interpreted as a recommendation to discard design theory and experience. This is, therefore, a convenient point to emphasize that consideration of alternative sample designs is essential. Many good texts exist, and surveyors will need to master appropriate sections of them or take professional advice.

A stratified, clustered sample is the almost universal choice for major surveys, so a few comments are merited.

The choice of strata and clusters is important and interrelated. In any national survey, stratification by region and, often, rural–urban categories is almost certain to impose itself. Although geographical contiguity does not always lead to the homogeneity needed for efficient strata, it can usually contribute something. Even if the sample size is not large enough to provide satisfactory district estimates, it will still be natural to build up national estimates from district figures. The possibilities of using different sampling fractions for different geographical strata should also be investigated, but this procedure is only likely to be worth adopting if there are gains on a number of items. It should be noted, however, that strata are not synonymous with domains of study—the parts of the universe for which separate estimates are required.

In multi-subject surveys there may be conflict between the different strata boundaries desired for different topics. As an example, it may be that a stratification by agro-ecological zone may reduce sampling errors for agricultural variables, but may not be efficient for demographic variables. Administrative problems may also occur if such zones cross district boundaries. We should want to see very good grounds for expecting substantial benefits from the scheme proposed before going against administrative simplicity.

Post-stratification may be worth consideration in such cases. In one sense any analysis which regroups data into interesting combinations provides a kind of post-stratification, although valid sampling errors may not be calculated in many of these cases.

Units within a stratum should be as like each other as possible, and strata should be differentiated as much as possible. This is because the variance between the strata 'drops out' from the sampling error, leaving only the sum of the total within stratum variances. Each stratum must be sampled. The guidelines for the division of the frame into clusters are different. First, as the aim of cluster sampling is to concentrate the survey into a limited number of groups, not all clusters are sampled. However, this concentration of the inquiry in a few clusters will not provide satisfactory estimates unless the individual sample units within each cluster are heterogeneous, covering the range of features found in the population as a whole. Often, units in a cluster are more like each other than they are to the rest of the population so that the inclusion of an extra unit in the sample from the same cluster contributes less to improving the accuracy of the estimate than would a unit chosen from outside the cluster. This similarity (or on the rare occasions when there is greater heterogeneity in the cluster than in the population at large, the dissimilarity) is measured by the intraclass correlation; and attempts are made to reduce it by making clusters heterogeneous. But if the stratum within which the clusters lie has been constructed to contain units as homogeneous as posssible, these objectives are in conflict. As Hansen *et al*, comment, 'PSUs [primary sample units] that are already made internally as heterogeneous as feasible [will provide] somewhat less opportunity for classifying such PSUs effectively so as to achieve substantial differences between the strata. But the object is to go as far as is practicable with this classification after the PSUs have been defined.'[18]

Natural clusters are unlikely to be equal in size; but if they are more than moderately variable the sampling error increases and there are other disadvantages. Further, clusters apparently of about the same size according to past information may turn out to have diverged by the time they are investigated. All clusters must be clearly delimited and there is usually no problem in grouping together the smaller clusters to form a larger one. There are, however, dangers in breaking down large clusters into smaller ones unless one is sure that the proposed smaller clusters are delineated properly and are identifiable in the field. This is another aspect of the danger of specialists at the centre too readily assuming that their changes can be applied in the field; or where, through the lapse of time, change of staff, and/or mislaying of documents, cluster boundaries decided during preliminary work are not applied properly when enumeration actually takes place. After considering these problems, Scott concludes: 'All these arguments tell in favour of a sampling method with clearly defined PSUs. While in many countries there are difficulties in defining PSU boundaries, it will generally be better to use such PSU sampling frames as are available than to rely on primary sampling units created in the field.[19] We agree. In general, then, it is dangerous to take risks over the clear definition of cluster boundaries, even if the clusters resulting from adherence to well-mapped boundaries are not as equal as one would like.

A second issue is the calculation of the optimum size of sample within a cluster or PSU. An approximate formula derived from a simple model given by Hansen is:

$$\text{optimum size} = \sqrt{\frac{C_1}{C_2} \frac{1 - \delta}{\delta}}$$

Where C_1 is the cost of adding a PSU to the sample, C_2 the cost of including an extra final unit, and δ is the intraclass correlation. If δ is low—say 0.01—then the optimal size will be approximately $10\sqrt{(C_1/C_2)}$. If δ is higher—say 0.1—then the optimum size will be $3\sqrt{(C_1/C_2)}$.

The cost functions assumed to estimate optimum sizes in this way are linear—that is, C_1 and C_2 do not change as alternative plans are considered; and there are no constraints—that is, it is assumed one can recruit incrementally at the appropriate level whatever the number of clusters and sampled units emerging from the calculations. These conditions do not normally apply in developing countries.

Where administrative, travel, and language difficulties are prominent, it is often easier to consider a limited number of feasible alternative schemes and evaluate their relative efficiency. If previous information about costs and intraclass correlations does not exist, then a range of values can be used to see how sensitive the rankings of the alternative schemes are within likely variations.

The manner in which the optimum size decreases as δ increases reflects the fact already mentioned, that when the cluster is homogeneous (a positive δ) the return from sampling extra units in the cluster is low. The relative inefficiency of cluster sampling designs is very large in some cases. Scott cites preparatory calculations for a Ghana survey for estimating the percentage unemployed, showing a cluster sample as being 20 to 30 times less efficient than a simple random sample in rural areas and 70 times less efficient in Accra. Comparisons of relative efficiency are usually made against the standard of a simple random sample, even though in many cases such a sample could never be a practical alternative. The measure is obtained by considering $1 + \delta(n - 1)$, when n is the number enumerated in the cluster. If $\delta = 0.05$ and $n = 10$, $1 + \delta(n - 1) = 1.45$, or the sample has to be 45 per cent larger than the simple random size. To estimate the size of a clustered sample required for a desired level of relative variability the formulae for simple random samples can be used, and the appropriate correction for specified values of δ and n calculated. Alternatively, if the size is fixed, an estimate of the relative variability of the cluster sample can be made by an adjustment in the opposite direction. The average cluster size expected can be taken for n.

Data on agricultural holdings, particularly crop yields, often exhibit very substantial intraclass correlations, of the order of 0.2 to 0.5. In such cases clustered sampling is extremely inefficient. Unfortunately, a popular type of survey involves an enumerator resident within the cluster undertaking objective measurements of holding-related variables. The number of sampled holdings in a cluster is chosen to provide the enumerator with a full work-load: 15 to 20 is a common range. The cost in sampling efficiency with this design is so high that the gain in data quality would need to be very great. Experience shows that the quality gain is often much less than anticipated; certainly less than is needed to justify the design. We will return to this problem in Chapter 13.

When serious defects of obsolescence and omission are thought to exist in the frame that has to be used, a design using 'compact clusters' may be suitable.[20] This involves enumerating the whole cluster selected (whether as a primary or secondary sample unit): this procedure may be attractive when the clusters are small—no preliminary listing and selection are required thus reducing time and costs if the extra labour of enumerating all units is small. There are also advantages for community studies and other investigations into social patterns.

Some sample designs are said to be 'self-weighting'. These are characterized by arranging probabilities of selection at different stages in such a way that the overall probability of selection of each final sampling unit is the same. Sample results can then be rated up straightforwardly by the reciprocal of the sampling fraction; there is no need to apply different weights to different parts of the data, and this is an advantage for multivariate analysis. In the case of a clustered design, sampling with probability proportional to size of the cluster followed by the selection of a fixed number of units within the cluster provides a self-weighted sample and also an equal work-load within each cluster—a considerable advantage in survey administration in many cases.

Samples are often chosen by systematic selection rather than by random choice of each sampling unit. A starting number is chosen at random, and then every subsequent sample unit is identified by adding a constant factor which is the reciprocal of the sampling fraction: for example, if a 5 per cent sample is being chosen, a number (k) between 1 and 20 (or 00 and 19) is chosen at random and the sample comprises the kth unit, then the $(k + 20)$th, $(k + 40)$th unit, and so on. (Note to beginners: *not* k, k + k, k + 2k.) If the population is arranged in the frame in a random order, then a systematic sample is equivalent to a random one. However, it is not often that units appear on a list in a random order; they are generally ordered in some way, often alphabetically, geographically, or by date of recording or registration. In this case a systematic selection provides in effect a rudimentary stratification of the sample. The systematic choice is easier than ordinary random selection, and can be executed very readily when the frame is laid out on a set of index cards, or on a serial list.

If the list exhibits a periodicity then there are dangers that a systematic sample may be badly biased. The effect depends on the relation between the length of the period, the sampling interval, and the length of the list being sampled. An example frequently cited is a sampling interval related to the arrangement of dwellings in a block so that a systematic sample contains all or no corner dwellings.

Surveys which involve 'before and after' comparisons or a series of observations over time raise special problems. One of the best summaries of current knowledge is given by Kish.[21] If one is primarily concerned with measuring change, it is usually desirable to maintain the same units in the sample. This allows full advantage to be taken of the correlation between the before and after observations which, when it is positive—as it normally is—reduces the variance of the estimate of the difference. When the survey is a periodic one generating time series data, a rotation design is generally used. This results in partial overlapping of units in successive samples. This retains some of the advantages of the positive correlation between observations on the same units, but spreads the burden over more respondents and diminishes the possibility of bias that can arise when frequently surveyed respondents cease to be representative. Replacing the sample also spreads the risk of selecting an unrepresentative sample on any one random selection.

4.6 SAMPLING AND NON-SAMPLING ERRORS

A crucial stage in planning a survey is deciding on the number of respondents or units to include in it. We have emphasized, in the Introduction, our belief that consideration of the non-sampling errors is at least as important in this decision as the level of the sampling error. From the derivation of sampling theory outlined in Chapter 1, it follows that the sampling error decreases as the size of sample increases. The relationship is a function of the square root of the sample size leading to the approximate conclusion that doubling the sample, n, will reduce the sampling error by a factor of $\sqrt{n}$. It is worth emphasizing that (except in a limiting case of a small population) the sampling error is not a function of the total size of the population, so the required sample is not a percentage of the population. It is surprising how frequently surveyors fail to grasp this point. The sampling error is, however, dependent on the variability of the characteristic of interest in the population or universe.

In any survey there are many potential sources of error that we cover by the term 'non-sampling errors'. One, an inaccurate frame, has been discussed in this chapter. There are the possibilities of non-response, erroneous response, observation or measurement mistakes, errors in recording or coding the information, and others.[22] Some of these may be accidental and may tend to cancel each other out; one respondent remembers a price as higher than the actual, and another remembers it as lower. But if the errors tend to be in one direction, e.g. a tendency of respondents to under-report income due to memory failure or suspicion regarding the purpose of the survey, then the estimate derived from the survey will be biased. Scott states 'studies have repeatedly shown the presence of alarmingly high levels of response error even on the simplest of survey questions'.[23] Measurement errors, notably measurements of crop yields, are also prone to substantial biases.

Different questionnaires, types of enumerator, or levels of supervision will give rise to different levels of bias: indeed, much of this book is devoted to advocating procedures which will, it is hoped, reduce bias.

The total survey error (σ_t) is the square root of the sum of the squares of the sampling and non-sampling errors, i.e.

$$\sigma_t = \sqrt{(\sigma_s^2 + \sigma_n^2)}$$

where σ_s^2 = the sampling variance

and σ_n = the standard deviation of the non-sampling error.

It is easily verifiable from this relationship that if σ_n is of the same order of size as σ_s, or bigger, reducing σ_s by increasing the sample will have only a marginal effect on σ_t. But a substantial increase in the sample size may have the effect of increasing the non-sampling error due to poorer quality enumeration and lower intensity of supervision. In such a case the resulting total error may be larger from the bigger sample than from the smaller.

It is not often possible to measure the non-sampling error, but if the surveyor believes that he can control it only by using a limited number of enumerators under intensive survey controls and supervision, then he would be wise to reduce the sample size to achieve this standard of enumeration.

If we knew more about bias, we might be able to argue that it will not be so important if our interest is in change over time, as the bias may remain constant from one survey to the next. At present, however, our fears of substantial and varying bias lead us to stress the importance of reducing it.

Notes

1. Cochran, W. G., *Sampling Techniques*, 3rd edition, Wiley, New York, 1977.
2. Brown, J., *et al.* (eds.), *Multi-Purpose Household Surveys in Developing Countries*, OECD, Paris, 1978.
3. Murthy, M. N., 'Evaluation of Multi-Subject Sample Surveys', *Int. Statist. Rev.* 42, 1974, p. 175.
4. Ainsworth, M., and Muñoz, J., *The Côte d'Ivoire Living Standards Survey: Design and Implementation*, LSMS Working Paper No. 26, World Bank, 1986.
5. Mahalanobis, P. C., 'Recent Experiments in Statistical Sampling in the Indian Statistical Unit', *JRSS* 109, 1946.
6. Deming, W. E., *Sample Design in Business Research*, Wiley, London, 1960, p. 9.
7. Hansen, M. H., Hurwitz, W. N., and Madow, W. G., *Sample Survey Methods and Theory*, vols I and II, Wiley, New York, 1953, p. 244.
8. Scott, C., *Sampling for Monitoring and Evaluation*, World Bank, Washington, D.C., 1985, p. 6.
9. Kannisto, V., 'On the Use of the Follow-up Method in Vital Statistics Sample Surveys', *International Population Conference, Liege, 1963*, IUSSP, 1964, vol. II, p. 395.
10. Hill, P., *Rural Hausa: A Village and a Setting*, Cambridge Univ. Press, Cambridge, England, 1972, p. 264.
11. Hill, P., *Population Prosperity and Poverty: Rural Kano, 1960 and 1970*, Cambridge Univ. Press, Cambridge, England, 1977, pp. 76–7.
12. Uganda Protectorate, *Uganda Census, 1959: African Population*, Entebbe, p. 10.
13. See, for example, de Graft-Johnson, K., 'The Statistical Problems of the African Census Programme', *Bull. Intl. Stat. Inst.* 46, 1975.
14. World Fertility Survey, *Manual on Sample Design*, Basic Documentation No. 3, 1975, p. 24.
15. Stevens, W. L., 'Estimation of the Brazilian Coffee Harvest by Sampling Survey', *JASA* 50, 1955, pp. 778–9.
16. Ibid., p. 779.
17. See Kish, L., *Survey Sampling*, Wiley, London, 1965.
18. Hansen, *et al.*, *Sample Survey Methods and Theory*, vol. I, p. 383.
19. Scott, C., 'Sampling for Demographic Morbidity Surveys in Africa', *Rev. Intl. Stat. Inst.*, 35, 1967, p. 165.
20. Zarkovic, S. S., and Krane, J., 'The Efficient Uses of Compact Cluster Sampling', *Bull. Intl. Stat. Inst.* 41, 1966.
21. Kish, op. cit.
22. O'Muircheartaigh, C. A., and Payne, C., *The Analysis of Survey Data: Exploring Data Structures*, vol. I, Wiley, London, 1977.
23. Scott, C., *Sampling for Monitoring and Evaluation*, p. 15.

5

The Case Study

'A singular case,' remarked Holmes.

The Sign of the Four

5.1 CHARACTERISTICS

The case study involves the detailed examination of a relatively few persons or items. The subjects of study must of course be appropriate to the matter in hand, but they will not usually be chosen by a formal sampling process. In some instances the subjects may be self-elected: for example, where the study starts with, or finally becomes dependent on, volunteer respondents. Or the study may just comprise all those persons with the relevant feature or features who come to the notice of the investigator.

These situations often arise in medical and psychological research from which the description 'case study' is taken. In social science 'the case study . . . is typically not of an individual but an organization or community'.[1] Once again, since many case studies are undertaken by single investigators as separate enquiries, the organizations or communities dealt with will generally not be selected by formal sampling.

The essential methodological feature of a case study is that it provides in-depth, detailed analysis. That is its strength, and it means that the enumeration is usually carried out by a professional, sometimes with a few qualified assistants. The method is indicated when it is necessary to probe deeply into the systems governing behaviour and the interrelationships between people and institutions; to establish and explain attitudes and beliefs, and to show why certain behaviour occurs. Case studies are particularly appropriate when a high analytical content is required, such as the study of causal relationships. They are free of the questionnaire and interview constraints of large sample surveys, and they are not limited by the implied urgency of rapid appraisal methods.

Because very few respondents, communities, or operating systems can be included in such a detailed probing study, they need to be selected in such a way as to ensure representation of the various types of interest in the study. If selected at random some types may be over-represented and others not represented at all. Stratification may resolve this problem, allowing for a random selection in each stratum. In many instances, however, deliberate selection of cases that clearly exhibit the character-istics of the types that they are intended to represent will be either forced on the researcher, because of the lack of a frame for stratification, or will be his chosen option. Some researchers employing the case study approach believe that random selection of the final unit and boosting the number of respondents are essential if their survey is to achieve 'respectability'. However, if the general structure of the survey is such that no statistically valid generalization of the results to a wider population can be made, these measures may be fruitless, and nullify the advantage of the case study method, if the increase in the number of interviews reduces the in-depth quality of the data.

The concentrated, skill-intensive nature of the case study enables the use of both objective methods of measurement and the detailed probing of attitudes and background. Unfortunately, the contrast between the nature and logical status of case study material and sample survey data is often drawn primarily to suggest that one method is better than the other. Such argument conceals the fact that many problems affect both methods, and distract attention from the main issue, which is, 'What method is most appropriate to the current stage of the enquiry given the resources available?'

The case study uses a mixture of methods: personal observation, which for some periods or events may develop into participation; the use of informants for current and historical data; straightforward interviewing; and the tracing and study of relevant documents and records from local and central government, travellers, etc. The approach can be very flexible and the progress of the study may often not be charted in advance, although this free-wheeling attitude will not be such a feature of commis-sioned project evaluation studies, pilot surveys, or medical and nutritional studies. The last more often resemble a controlled trial in natural, not laboratory, conditions.

One consequence of this flexibility is that a case study is entirely dependent on the ability, the experience, and the ingenuity of the investigator, who needs to be able to observe, interview, record, and continuously review the material collected. Considerable mental and physical stamina are required. In some ways a case study is inexpensive because of the small numbers involved. The demands on the investigator are, however, great: in particular the need to maintain an effective rapport with the group at all times, often over a long period with continual pressure; and to record as soon as possible after the event data which cannot be noted at the time they were obtained, since the act of recording could affect the source.

Another consequence of the open-ended nature of field studies is the part played by chance events. The course of an inquiry may be deter-mined by the discovery of some unexploited documentary source, or by an

encounter with a particular individual. Two quotations from Hill illustrate the situation. First: 'chance plays such a large part in research of this
kind, that it is usually difficult to lay down detailed plans in advance. Just
as I had not thought two weeks in Northern Ghana would be so
rewarding, so I was astonished at the need to devote as long as $3^1/2$ years
to studying the migrant cocoa farmers. But if one cannot plan one's work
in advance, how should it be directed? For myself I depend very much on
my naïve feelings of *surprise*—holding that the most surprising "events"
are most worth pursuit.' Secondly: 'I made an expedition to Anlo country,
in south-eastern Ghana, to study the famous shallot-growing industry
there. . . . In the event I was diverted by the surprise I felt over the
elaborate organization . . . of the seine-fishing companies.'[2]

   The flexibility of the case study method makes it particularly suitable in
the early stages of an inquiry. In the framework of the general survey
method, the pilot survey is a case study. More generally, a case study can
explore a field or situation, refining or ruling out preliminary hypotheses
and suggesting new ones, and possibly providing orders of magnitude of
key features. The case study can also be effective at later stages as special
groups or patterns that have been found during the analysis of some large-
scale survey can be made the focus of a detailed inquiry. The case study is
also essential for those topics where the detail and precision of the
information required are such that only a very small number of cases can
be investigated at any one time: for example, nutritional studies of intake,
detailed study of household transactions, or of local market channels.

   The differing requirements and appropriate subjects of case studies and
of larger-scale investigations show there is considerable scope for effective
co-operation between official statisticians and individual academic workers in the universities, in research institutes, or accredited to governments
under some project. By concentration on what each can do best, the final
result is richer. Academics have the satisfaction of seeing at least part of
their work fitting into a wider framework and helping to influence policy
for the benefit of those who have exposed themselves for study. Official
statisticians can take advantage of trained assistance to cover areas or
issues which they could not cover because of lack of resources or specialist
knowledge.

   Such a collaboration is not easy to arrange and bring to fruition. There
may be problems of timing, particularly when outside experts are
involved. Two issues of principle are also likely to be difficult to resolve.
Academics will be conscious of their need to retain independence and
scope for critical discussion of policy and its effects, when it seems this is
needed. Official statisticians will be concerned with problems of confidentiality, and with the possibility that the academic working in a more
flexible open-ended way may not keep to the timetable imposed by
government needs, even though a linked study will not be so subject to the

element of chance, expressed by Hill. Clearly good personal relations and mutual confidence will be important in an effort to achieve such a collaboration; but it is very rewarding when it is successful. For example, a general market survey mounted by the appropriate ministry with FAO assistance was supplemented by special studies by two economists of potato marketing and maize marketing. Everyone benefited.

The limitations of case studies need to be recognized. The main problem is that of generalizing the findings of the study. The first level of generalization is, in a sense, within the study itself; what is the status of data about the whole village or system obtained from purposively selected respondents? How can we generalize from the statements of witnesses not randomly selected to valid statements about the whole group under study?

If a population is completely homogeneous, a sample of one will give information from which statements about the whole population can be made. No set of natural objects or beings is ever completely homogeneous, but experience has shown that many groups are sufficiently homogeneous in some respects for limited inquiries to work satisfactorily. This view is summed up by Lazarsfeld and Barton: when one is dealing with 'groups with a nearly homogeneous culture, in which one set of prescribed rules is just about universally carried out by the population, it may require only the observation and interviewing of a relatively few cases to establish the whole pattern. The same argument can be applied to studies of a quite homogeneous sub-culture.'[3] Experience and judgement are needed to decide whether the conditions of homogeneity are in fact met when it is assumed they are. The general criterion will be consistency in the testimony of the informants, with conscious checks to test that circumstances which could lead to a misleading consistency are not present. For example, if all informants are from one small group or are linked in some way, they may consciously or unconsciously provide a consistent but incorrect picture.

The second level of generalization is the ability to generalize from the study of a single, or a few, respondents, communities, or villages to a project area, a region, or even a country. Such studies based on purposive selections, or random selections of insufficient size, cannot be used to make valid inferences about the incidence of a phenomenon or the average value of a variable in a wider population. But, although one may not be able to generalize from a case study, one may be able to reject existing generalizations. A number of case studies have certainly been important, for example, in undermining widely held views about the irrationality of smallholder economic activities. Many development interventions are postulated on the assumption that the targeted population, if offered a particular service or stimulus, will respond with a fundamental change in their behaviour. A case study conducted when such an intervention is at the preparation stage may reveal that the constraints to such a behavioural change have a different basis.

We have referred to the key role played by the primary researcher in any case study, and should include a reference to the difficulty, often referred to as the 'personal equation' of the observer. Thus investigators may be affected so that they no longer remain objective observers and recorders. They may overidentify with their subjects and their view of what is going on, or their lines of enquiry may be constrained too much by the relations that develop between them and the people they are studying. They may begin to adopt the unconscious assumptions of their subjects of study. On the other hand, they may miss or misinterpret what is happening by failing to achieve a proper rapport or by importing inappropriate interpretations from other cultures. Most field workers are of course very conscious of these possibilities; but the pressure of the work, possibly combined with long periods of isolation from colleagues or from other contacts outside the area under study, may gradually dull or narrow their responses.

The case study is a valuable method of enquiry, which may be particularly appropriate in the context of a wider investigation. It requires great effort by an investigator of all-round ability and specialist knowledge. One of its special values is that it frequently shows the limitations of conventional wisdom, particularly incorrect stereotypes of rural life and activities which have often affected development policies in the past. The method also provides an effective way in which academics and officials can work together to improve the framework for policies for change and development. Resources for high-level investigations are so short in developing countries that strenuous efforts need to be made to co-ordinate them.

For the developing countries, the problems are those of decision, not of generalization for science. The question is: 'Are development policies likely to be more effective if they take into account the case study findings, than if they do not?' Experience suggests the answer is 'Yes', although this does not, of course, imply that individual results do not need sceptical scrutiny.

We give, in the remainder of this chapter, some examples of methods for conducting enquiries that fall within our general definition of case studies.

5.2 PARTICIPANT OBSERVATION

Developing countries have provided the subject matter for numerous anthropological case studies of single tribes or communities. This classical type of study is giving way to a similar design that involves social scientists who share a common cultural inheritance with the community of interest and who participate in the life of the community in order to be observers from the inside. Hence, the name of this type of case study, participant observation.

The basic method involves the researcher living within a community, in identical conditions to those of the community members, and joining in the normal economic and social life of the community for an extended period of at least several months. Information is obtained through normal conversation and the use of a trained eye, and requires that the community accept the observer as an 'insider' and trust him with its confidences.

This type of case study is particularly useful when there is a need to understand the attitudes, perceptions, and constraints influencing a community in a rapidly changing environment. It is, therefore, growing in popularity as a means of qualitative evaluation of a development intervention. One practitioner, Salmen, writes:

Coming from a group of people who are developers to join another group who are undergoing the process of development one learns how far the two often are from each other and how they may be drawn closer together. Wherever I or others looked at projects from the ground up, whether in housing projects in Thailand, fishing co-operatives in Brazil or agricultural endeavours in Bolivia, we saw areas of concern which transcended sectoral and regional differences. We learned that many project beneficiaries simply do not understand the nature of the project from which they are to benefit. Often there are established local interests, political leaders or entrepreneurs, whose influence . . . has been ignored or underestimated. Often projects are seen in isolation from the context in which they are placed, as abstractions rather than interfaces with people who have unique histories, location and cultural composition. . . . How a project touches the inner core of a beneficiary and becomes a catalyst for self-improvement comes to light as one listens to the people.[4]

The following brief guidelines are drawn from those offered by Salmen.
(a) Prepare in advance by a study of the history of the community, and identify a preliminary agenda of issues which it is intended to explore.
(b) Establish a residence in the community for at least two periods of several months.
(c) Gain the trust of the community by openness and frankness; participate in community activities whilst retaining an independent stance on local controversies and disputes.
(d) Get to know the community by observation, particularly how its organizations actually work. Discuss key issues with leaders, activists, and community members only after acquiring this knowledge and achieving their trust.
(e) Consult community leaders regarding the focus of the report.

5.3 LONGITUDINAL STUDIES WITH DETAILED RECORDING

Case studies of the type just described focus on achieving a measure of understanding that enables the observer to interpret a situation in a qualitative manner. Other case studies have the objective of collecting data with a high order of accuracy to allow quantitative relationships to be formulated; for example, dietary studies of children prone to malnutrition require careful observation and measurement of the children over an extended period. Accuracy of data takes precedence over sample size, thus indicating a case study approach.

Farm management studies are a popular example of this type. Typically such a study requires an investigator to maintain regular, almost continuous, contact with a few farmers throughout, at least, one agricultural year and to record in detail the scale of farming operations, the volume and costs of inputs, including hired and family labour inputs, the production achieved and its disposal, and changes to farm assets, such as land and livestock acquisition and disposal.

Maxwell stresses the advantages of the case study approach: 'Whatever the number of farms in a particular category, case study programmes should have an overall limit of around ten. There are sharply diminishing returns to larger numbers because of less intensive or less frequent contact and sharply increased costs because of increased time for both data collection and data processing.'[5]

This position is not supported by all. One advocate of farm management studies believes that larger samples are essential. 'These analyses are quite complex and the inherent small sample size of an informal attempt can leave too much room for extreme bias in the results.'[6]

But experience, including that of the project referred to in this quotation, shows that the cost of farm management studies on large samples, the difficulty in processing large volumes of data from multiple visits to each farm, and the likely data quality problems when large enumerator forces are used are major arguments in favour of the limited case study approach of Maxwell.

Even when the number of cases is small the complete recording of all farm system variables may lead to indigestible volumes of data as well as taxing the patience of the farmers. Maxwell, in drawing attention to this, makes another point that provides the linkage with more qualitative case studies: 'Often the most important aspect of a case study will not be the formal data collected but the understanding of management practices derived from conversations with the farmer.[7]

5.4 DIAGNOSTIC STUDIES

In the management of developmental activities, particularly in the context of specific project preparation and implementation, there is a frequent need to carry out a quick diagnostic study of a problem that is identified as a constraint to the acceptance of the project or activity by the intended participating population. Speed in execution and flexibility of approach are crucial to such a study. The methodology used will depend on the type of problem to be diagnosed, but often a quick survey in one or two localities that exhibit in a marked manner the phenomenon to be investigated may be the only feasible option. Two versions of such an approach used in rural areas form part of the procedures described as Applied Farm Systems Research.

CIMMYT agricultural economists, notably Collinson and Byerlee, use a three-stage procedure for their studies of farming systems. One of these is an exploratory survey in which a multi-disciplinary team spend some time in a locality engaging in unstructured interviews with farmers and other community members, using a check-list to focus their own probes during the interview:

the interviews should be carried out in the farmer's field in order to relate questions to observations in the field. Interviews should be conducted in a relaxed manner. Use of pencil and paper should be avoided. . . . It is possible to cover only a part of the . . . checklist in one interview. . . . What information is included will depend on what practices a farmer is following, what problems he is experiencing . . . much can be gained—particularly in interviews with traditional leaders—by discussing practices and variations commonly followed by farmers in the area. . . . After each day's work, it is useful for the researchers to evaluate what they have learned, formulate new hypotheses and determine . . . the key gaps . . . in their understanding which should be explored in further interviews.[8]

A similar version, developed in Latin America, goes under the name of a 'Sondeo Survey'.[9]

The CIMMYT approach leads from the exploratory study to a formal sample survey using a structured questionnaire that is designed in accordance with the findings of the field study. This is an interesting example of the useful linkage between case studies and sample surveys that we referred to in the opening section of this chapter.

5.5 COMMUNITY STUDY

It is not always necessary to install a participant observer into a community in order to collect community-level information that may be required. For a study of a single community it may be possible to use existing records, supplemented by a community questionnaire and, in certain cases, a census of the households in the community.

An example where such an approach may be both feasible and necessary is provided by the involuntary resettlement of a village or community due to a development intervention, such as the construction of a dam, or as a consequence of an environmental disturbance. In such a case, records are likely to be maintained by health and social service agencies—these can provide much of what is needed for monitoring the community's efforts at establishing itself in a new location. A census within the community might well be necessary, and a community questionnaire applied to community leaders and prominent members should fill in the gaps regarding availability and use of community-level resources and facilities.

Community questionnaires are now used in a wide range of inquiries.[10] They are economical, since the questions need to be asked of only a few people, and certain information can be obtained with reasonable reliability for small communities of up to, say, 2,500 inhabitants. The existence and location of schools, markets, health centres, and service offices are not likely to be in dispute, and the existence and types of roads, water supplies, sanitary facilities, and fuels will be ascertainable. Information about the distribution of, and access to, some facilities may be inadequately reported and possibly biased.

The community questionnaire is one where the rule against the 'since we're there, we'll ask this' practice may be relaxed, since the costs involved are so low. If the information on some aspects turns out to be rather poor, it may nevertheless be of interest to note what the respondents involved reported.

Group interviews may also play a part in such a case study, and also in the diagnostic studies discussed above. Obtaining the reaction of a group provides the opportunity for the researcher to note the interaction between the group members as they respond to the issues raised. Group meetings are more likely to succeed in communities that are accustomed to discussing local issues in such a setting. The researcher needs to be aware of the dangers of a few individuals dominating the discussion and appearing to reflect a group unanimity that is deceptive. Silence from group participants does not necessarily mean consent, but may indicate politeness or diffidence.

Notes

1. 'Observation', *International Encyclopedia of the Social Sciences*, vol. xi.
2. Hill, P., *Studies in Rural Capitalism in West Africa*, Cambridge Univ. Press, 1970, pp. xiii–xiv.
3. Barton, A. H., and Lazarsfeld, P. F., 'Some Functions of Qualitative Analysis in Social Research', *Frankfurter Betrage zur Sociologie* 1, 1955, reprinted in McCall, G. J., and Simmons, J. L. (eds.), *Issues in Participant Observation*, Addison-Wesley, Reading, Mass., 1969, p. 82.
4. Internal Communication by L. Salmen to Author, 1985.
5. Maxwell, S., 'The Role of Case Studies in Farming Systems Research', *Agricultural Administration*, 21, 1986, p. 156.
6. Swanberg, K., reported in Deboeck, G., and Rubin, D. (eds.), *Selected Case Studies on Monitoring and Evaluation of Rural Development Projects: Eastern Africa*, vol. i, World Bank, Washington, DC, 1980, p. 12.

7. Maxwell, op. cit., p. 157.
8. Byerlee, D., Collinson, M., *et al.*, *Planning Technologies Appropriate to Farmers: Concepts and Procedures*, CIMMYT, Mexico, 1980.
9. Hildebrand, P., 'Combining Disciplines in Rapid Appraisal: The Sondeo Approach', *Agricultural Administration*, 6, 1982, pp. 423–32.
10. Ashe, Jeffrey, *Assessing Rural Needs: A Manual for Practitioners*, Volunteers in Technical Assistance, Washington, DC, 1979.

6

The Questionnaire

'What the devil do you want here?' 'Ten minutes talk with you,
my good sir,' said Holmes, in the sweetest of voices.

Silver Blaze

6.1 GENERAL

We set out first our usage of 'questionnaire'. The *Dictionary of Statistical Terms*[1] defines it as a 'group or sequence of questions designed to elicit information upon a subject, or sequence of subjects, from an informant'. A sheet with a tabular layout, on which appropriate figures are entered to record replies to questions stated or implied by row and column headings, is often referred to as a 'form'. We do not find this distinction between a questionnaire and a form particularly useful in developing countries since so many of the records of responses are a mixture consisting of both questions and tabular layouts. Neither shall we use the term 'schedule' or 'instrument', which occur in some reports.

We shall be mainly concerned with questionnaires used by an enumerator in an interview with a respondent. We cite some illustrative examples but cannot reproduce them in full. The design of this kind of questionnaire is, in some respects, simplified, since it is an 'in-house' document and the explanations of concepts and discussions of procedure can be given to enumerators during their training course. An unthinking exploitation of this advantage, however, can lead to a cluttered, badly laid out appearance; the designer has relaxed his standards because the form will not be filled in by, or shown to, respondents. Such a neglect of the enumerator's wants will almost certainly lead to poor recording.

Example: An eight-page questionnaire on income and expenditure required the enumerator to complete twenty-one crowded tables of different designs, layout, and format (Southern Africa).

Many apparently minor, but actually important, details require concern for the enumerator's convenience, and more than repay the time and effort taken by leading to better quality processing and data security. They include:
(a) good quality paper and printing;
(b) appropriate size of the questionnaire;
(c) printing on one side only;
(d) standard headings (what printers call a 'house style').

Financial limits may be strict, but a cyclostyled questionnaire produced on cheap absorbent paper, with poor typing, smudged headings, sloping lines, or dots all over the place, will affect the quality of the enumerator's work. He is expected to produce results of a high standard in arduous conditions, and can legitimately expect that some attention is paid to his convenience and that his working materials are of reasonable quality.

Some questionnaires extend over an apparently never-ending sheet that almost turns into a scroll. A foolscap or A-4 size sheet, or some similar size that, fully opened, matches the dimensions of the clipboard issued to the enumerator, is the maximum he should be expected to tolerate. Printing should be on one side of the paper only.

Example: A single-sheet questionnaire, 50 × 30 cm., for a household survey (Caribbean, 1977). This would have been easier to use if printed as a double sheet (single sides only).

The heading, including the name of the survey, year, executing agency, and country should be clearly shown at the head of the questionnaire. Also at the top of the sheet, and ruled off from the main body of the questionnaire, should be the page number of the sheet in the set, and the locational information, including codes, spaces for the enumerator's name, date of interview, and the supervisor's signature. Individual sheets of a questionnaire have an unfortunate habit of becoming detached from the remainder of the set during the survey, particularly during transportation from field to office. If the identification code is not present on each sheet this may cause delay at the data editing and processing stage. It may be advisable to show the locational information in words as well as in code. The casual error rate made by the enumerator in recording codes will be proportional to the boredom caused by repetition. A code such as 1400726 if written on several score sheets will too frequently appear as 4100276, 1400276, or some such version. Such simple errors can cause serious delays later on.

There is no doubt that questionnaire design is often the worst executed stage of survey preparations. Lengthy reproductions would be necessary to document this fully, and provide extended guidance. The extracts and examples we give will, we hope, help to create a greater awareness of what it is practicable to expect of an enumerator and to ask of a respondent. Many actual questionnaires would never have been passed for reproduction if the surveyor had tried out his demands on a colleague or on himself. Another danger is the inappropriate adoption of a model designed for one purpose in an entirely different context. One questionnaire widely available in developing countries is that recommended and used in the World Fertility Survey. Its success in its own field, with its verbatim questions, skip instructions, and stage directions to enumerators, does not mean that such a model is appropriate for a budget or agricultural survey.

Questionnaires must be tested in the field. This is normally done in the course of a pilot survey. The vocabulary used frequently means little to the respondent, particularly in agricultural surveys. 'Holdings', 'assets', 'remittances', 'wholesale', 'born alive', 'here', 'present', will require explanation. 'Normally' or 'usually' provide opportunities for great variation in response. The sophistication of the respondent, the level of enumerator, and the wording of the questions must be matched to each other.

6.2 THE PRACTICAL QUESTIONNAIRE

The practical questionnaire has to implement the results of the assessment we discussed in Chapter 2 of 'What is the minimum we need to know?' since, by the time the questionnaire is literally on the drawing-board, it is too late to query the basic objectives of the survey. However, it is not too late to examine again the need for each specific item of data suggested. The surveyor needs to keep the balance between 'How few questions will suffice?' and 'How many questions dare we ask?'

A glance at many questionnaires is sufficient to show that controls have not been applied. The tendency, as explained in Chapter 2, is to play safe—to include items that may be useful rather than run the risk that it will be discovered, too late, that an omitted variable may explain the relationship under study. This approach results in a questionnaire that includes all the questions that wit can devise, rather than the carefully distilled minimum list needed to achieve the main objectives; although sometimes, of course, even the minimum list itself is substantial. The World Fertility Individual Core Questionnaire[2] is not as forbidding as it seems at first sight, owing to the skip instructions; but any one respondent may be asked approximately 100 questions. What are we to make of a questionnaire for an agricultural survey carried out in Latin America in 1978 involving 40 pages and over 250 questions to a farmer, many about matters to which he could not be expected to have a ready answer?

Accuracy of response generally declines with length of interview. Interviews that extend beyond one hour may well cause fatigue to both the enumerator and respondent; and the more formal the questionnaire, in the sense that the enumerator must stick to a prescribed wording, the greater the risk of this happening. We recommend one hour as the desirable limit on any one single interview, and it is clear that this limit affects the number of questions that can be put. Nor is it merely a matter of length. The greater the depth of the questioning—the more the respondent is expected to think deeply about his attitudes, or to dig into his memory—the more limited should be the field of study. It may be reasonable to probe deeply actions and motivations regarding one aspect of behaviour, but excessive immediately to submit the respondent to a similar examination of another.

One example of an integrated questionnaire to measure living standards, which defies the recommendations just made, has been recently tested in Africa and Latin America. The book-length questionnaire covers all aspects of living standards such as demography, fertility, nutrition, employment and labour, income and expenditure, social conditions, and agricultural practices. All household members are interviewed and/or measured, on two or more separate occasions.[3] There are features of these tests in the context of survey management that provide valuable lessons (see Chapters 7 and 8), but we have grave doubts regarding the length and coverage of the questionnaire. At the least, the pilot surveys reveal that respondents will put up with such extended interviews; the issues remain whether they should have such demands placed on them, and whether there will be a consistent quality of response over the entire interview period.

One aid to discipline in questionnaire content is the preparation of a tabulation programme in advance of the finalization of the questionnaire (as discussed in Chapters 2 and 9). The 'play safe' syndrome must be resisted. When the draft questionnaire has been prepared it must be edited, ordered, and pruned in the way written work is treated.

The type of questionnaire required is, of course, dependent on the methods to be used to collect the data. For example, if crop areas are to be obtained by interview, it may not be possible to obtain the areas of individual plots, only the total area under a crop. Even the latter may be feasible only in approximate rounded terms—perhaps to the nearest hectare. On the other hand, if the plots are to be measured by the enumerator, the questionnaire design must provide space for the details of the calculations and the results, on a plot-by-plot basis to a fairly high degree of accuracy.

6.3 THE RESPONDENT'S SITUATION

There are many ways of conducting an interview with a respondent. The approach can be conversational, inquisitorial, programmed; the sequence can be haphazard or strictly structured; the questions precise and closed, or broad and open ended. Whatever the approach, the following questions must first be answered by the surveyor:

(a) Will the respondent understand the question?
(b) Will the respondent know the answer?
(c) Will the respondent reveal the correct answer?

Only if the surveyor is satisfied that the answers to all these questions are 'yes' should the proposed question appear on the questionnaire: otherwise the questionnaire cannot be expected to succeed. Consider the following common question: 'State current income' (or a similar wording). Will the respondent know what is meant by income? Is it family income? Are gifts and gratuities to be included? What about income in kind rather than cash? There are scores of definitions of income—which one will the respondent adopt? Almost certainly the definition will be a simple one that differs from the survey requirement; and there will be variation in the definition adopted by different respondents.

Another complication is that many questions are worded, or expressed, in such a way that the respondent consciously or subconsciously gets an impression that certain answers have been foreseen, even expected. Depending on the type and mood of the respondent, it.is quite possible that, in such a case, he will give what he considers to be the expected answer. This is particularly likely with open-ended and inquisitorial questions that start 'Why do you . . . ?' or 'What is it that . . . ?' At a training course for statistical officers from various African countries the participants were divided into three groups for a pilot study in a nearby village. One of the questions, taken from a questionnaire in current use, asked for the three most important reasons why the farmer did not expand his agricultural operation. Each group was instructed separately, but with identical instructions specifying the need to avoid leading the respondent to an answer. Do not mention possible reasons, they were told; do not give examples. Deliberately, however, the instructor casually mentioned one set of three likely reasons to each group: shortage of land, labour and tools; lack of money, seed, and fertilizer; or the poor quality of soil and the lack of roads and markets. Each group interviewed a random selection of farmers in a nearby village, and the most commonly stated set of constraints in the responses for each group were identical with that mentioned casually to the interviewers by the instructor!

6.4 THE RECALL PERIOD AND THE UNKNOWN ANSWER

The second and third test questions given in the previous section relate to the respondent's knowledge of the answer and his willingness to reveal it. These are difficult topics. Many surveys are built around questions the respondent cannot possibly answer. In some respects this betrays the wish to get a lot at little cost: a big sample enumerated in a single visit on a multiple of subjects. As surveyors, we often deceive ourselves, and the users, that the required data can be collected by means of a simple interview when this is not so. As a result the data collected are of poor quality; and the respondents have been subjected to an interview that they should have been spared.

Much of the discussion regarding the biases in data obtained by interview is centred around the question of the optimum recall period to use for data on various topics. We suggest there is rather more to it than that. The commonest error is to ask the respondent a question that it is impossible to answer accurately. This may be because either the answer was never known or too much is expected of the respondent's memory. A further factor affecting the quality of the response is that the respondent may be reluctant to reveal the answer. It is unreasonable to ask most respondents in developing countries for crop areas or milk yields, for net income from an informal business, or for the weight of food consumed. Their land has never been measured; the cows are milked by a herdsboy some distance from the farm; account books are not kept; nobody, except the affluent on a slimming course, keeps a record of the weight of food consumed. Such mistakes are often due to the transfer of inappropriate methodology. In some developed countries the housewife buys much of her food by weight or by standard units, for example, tins, packeted cereals, a standard loaf of bread. She may be able to record the household consumption in these terms. In developing countries the diet of most of the population will consist mainly of staple foods, unpackaged, sold in variable units, or produced and harvested from their own plots for home consumption with no records kept of weight or volume.

The length of the recall period is irrelevant in these cases, since the answer is not only unknown now—it never was known. If a long recall period is used the respondent, instead of attempting the impossible, will reply with a figure that he believes approximates an average or normal level. The response error is not due to a memory lapse, but to an error in guessing the value of this average, and ignorance about any variation from the average for the period under reference. A question about expenditure on food over one year may lead to a response produced by assuming a monthly average and multiplying by twelve.

Example: In a survey of expenditure using a one-year recall, figures recorded for one household were exactly divisible by 12 for 80 per cent of the entries made (Latin America, 1975–6).

A properly framed question on food *expenditure*, but with a recall period of one day will present the respondent with a true memory recall task. He will recall the actual purchases made and an estimate of their quantity with a greater or less degree of accuracy. These data will exhibit much greater variation about a mean for, in addition to the error in the estimates, the true daily variation in food expenditure by a household is high, since purchases may range from zero to a figure that on pay day may be as high as the total for the preceding week.

If the question is such that the respondent is expected to remember each event, together with associated particulars, the weight of evidence is that 'the level of reporting errors varies directly with the period of recall'.[4] A guiding principle, therefore, must be to keep the recall period as short as possible, bearing in mind the following considerations. The ideal recall period will vary according to the significance of the item being recalled and its frequency of occurrence. Income of a salaried individual cannot be collected with a reference period of less than one month. Recall of expenditure on major items, such as bicycles, radios, or furniture, must cover a relatively long period because the frequency of purchase of such items is low; use of a short reference period such as one week or even one month, will result in many responses of zero and a few of high values. The quality of the mean estimated from such data will be poor unless the sample is very large.

As mentioned earlier, a very short recall period, such as one day or three days, may be appropriate for the recording of consumption patterns, but will give rise to high variation in the data. Moreover, the data collected in the first visits may well have to be ignored since there is considerable evidence that the reported consumption will be untypical because of the so-called 'prestige effect', that is, a desire of the respondents to create an impression that their way of life is better than it really is. Experience shows that, with good enumerators, this type of response error disappears fairly quickly. Towards the end of the survey the repeated visits to the same household may cause respondent fatigue leading to a deterioration in the quality of reporting.

The reference period that the respondent is asked to recall should be closed, that is, the start and finish points should be clear in the respondent's memory. 'Yesterday' is an example of a closed reference period, whereas to enquire in the middle of September about sales during August is to use an open reference period—the respondent may not remember whether a particular sale occurred at the end of July or 1 August, or whether another sale was on 31 August rather than 1 September (unless the dates happen to coincide with a weekly market day). If dated records of sales are kept and consulted then the reference period would be closed for those recorded sales. One end of the reference period will certainly be closed if it ends at the moment of the interview. Open-ended reference periods lead to errors of recall in which events are improperly transferred into or out of the true reference period—the so-called telescoping and end effects.

There is some controversy in the literature over the differential effects of varying recall periods. The quotation from Raj above, that the quality of report declines with the period of recall, is in line with psychological opinion and common sense. He cites results supporting this view. On the other hand, Zarkovic[5] describes two studies that seem to suggest that expenditure or consumption may be better recalled over one month or even one year, than over one day or one week. One problem is that the 'true' figures by which to assess the different reporting periods are not known. As we have already pointed out, the figure for the longer term periods may well not be obtained by the same process of mind as those for much shorter periods—a rated-up average being used for the former and actual detailed recall for the latter. If we restrict ourselves to processes involving specific recall of individual items, our experience is in line with common sense expectation and the evidence cited by Raj. We suggest that the surveyor should consider the following issues when he is reviewing the pros and cons of interview or measurements, or of short and long recall periods.

(a) Was the answer ever known to the respondent? If not, neither a long nor short recall period is of use. Measurement, and in some cases longitudinal observation, may have to take the place of an interview.

(b) How serious is memory decline likely to be? Some events make a deep impression. A woman will remember bearing a child that lives, but she may not be able to fix the date of birth. A different form of question may improve response. Thus Blacker obtained a substantial improvement when a single question on number of children ever borne was replaced by three separate questions on the number still living with the mother, the number living elsewhere, and the number who have died.[6] For frequent events in the everyday circumstances of life a short recall period is desirable.

(c) Will the data from the use of a short recall period exhibit such high variances that the resulting estimates of means are unreliable? If so, may not a crude average estimated by the respondent over a period of one month or one year serve the purpose of the surveyor better? This crude average will reveal little about variations over time and may be grossly biased; it will, however, be much cheaper to collect.

The design of the question must also take account of the possibility that the respondent will be reluctant to give the answer to the question even when it is known. This reluctance may be due to a variety of reasons: bad enumerator·—respondent relationship; fears about repercussions when the facts are known, for example, increased taxation; and traditional beliefs that rule out the numbering or discussion of certain events. There is an extensive literature on these topics, particularly with regard to child and mortality data and to data on income (including livestock in traditional pastoral societies). The solution may be found in the use of alternative questions that will allow the required data to be estimated (for example, questions about orphanhood as a basis for estimation of mortality), in the use of very well-trained and skilled enumerators, or in the substitution of direct counts or measurements, not necessarily related to any particular individual or household.

6.5 THE HOUSEHOLD QUESTIONNAIRE

The commonest type of questionnaire, or part of a questionnaire set, is the single sheet on which each individual in a household is listed row by row, and particulars about each individual are recorded under various column headings. Population census questionnaires are frequently of this type, and most surveys using a household as the sample unit include such a sheet in the questionnaire. Figure 6.1 shows an example.

The main issues in designing such a questionnaire are:
(a) column headings and widths;
(b) pre-coding, whether completely or partially;
(c) the sequence of questions;
(d) the appropriate choice of questions.

The popularity of the single-sheet questionnaire in census and large surveys needs no explanation. The convenience carries with it a major problem—the lack of space for fully written out questions, instructions, and codes. Column headings such as those shown are both short and unambiguous and present no problem. The codes for male or female, and single, married, divorced, and widowed are either self-explanatory or easily explained.

Sex	Age	Marital Status
M/F	Years	S/M/D/W

A question such as the following (which appeared on a 1978 question-naire) required nearly 5 cm. of vertical space with one or, at most, two words on each line, and was difficult to read: 'If you neither worked nor looked for work last week, when did you last look for work? (See below for codes.)' Such a long question, printed in full as it is intended to be used by the enumerator, is unsuitable for this style of questionnaire, and is not in accord with the rule given above that headings should be short and unambiguous. More important, column widths should be varied primarily according to the demands of the *answer*; the printing of questions in full results in column widths varying according to the length of question. Thus columns requiring only a Yes or No answer may be 2 or more cm. wide, whereas names, etc., are often allowed only 1 cm. If the question is long but the answer is of two-digit length, a much abbreviated or even coded heading should be used, with the question in expanded form as a footnote. Footnotes are also preferable for the details of the codes to be used, unless they are as simple as in the two examples shown above. In extreme cases, the column heading, with a question spelt out in full and all possible codes specified in it, can occupy more vertical space than that allocated to the rows on which the data are to be recorded. This is unbalanced in appearance; and the numbers of rows may be reduced so that two sheets may be needed for one household. The unit in which the answer is to be

KENYA GOVERNMENT: CENTRAL BUREAU OF STATISTICS: RURAL SURVEY 1976–77   CONFIDENTIAL

FORM 1: HOUSEHOLD MEMBERS

Province —————— — District

Location —————— Date

Code

Card No.   0   1

Total in H'hold

Serial Number	Name	Relationship to Head of Household	Age 00 if below 1 yr.	Sex 1 Male 2 Female	At School Full-time 1 Yes 2 No	Works on Holding of Household 1 Yes 2 No	Main Occupation	Other Occupation	Normally Present 1 Yes 2 No

FIG. 6.1. *Source:* Central Bureau of Statistics, Republic of Kenya, *National Integrated Sample Survey Programme (Questionnaires 1974–8)*, Nairobi, 1979.

recorded must be specified in the column heading. Kilos or grams, acres or hectares, cents or shillings; these cannot be relegated to footnotes.

Pre-coding, where feasible, is recommended. Two possible methods, in which a male has been coded on the first row of the questionnaire and a female on the second, are shown.

(a)		(b)	
Sex		**Sex**	
1	2	M 1	
(M)	F	F 2	
M	(F)	1	
		2	

If the entries are numeric, and they should generally be so with pre-coded questions, an individual 'box' should be provided for each digit: this facilitates data entry on a computer file.

The size of the boxes for recording numbers should be adequate for the enumerator (standing, perhaps, in the open air with no firm base to press on) to write a number clearly and completely within the box. The tendency to write numbers that look like other numbers will be increased if the box is too small. In this context, Arabic numbers can cause particular difficulty, for example, a zero looking like a full stop. A standard method of correcting wrong entries should be taught to enumerators in their training course. A single or double deletion stroke is usually effective.

The sequence of questions is as important in a columnar type of questionnaire as in a listing of questions considered in the next section. Skip instructions are difficult to handle with a succession of columns, and may lead to answers being entered in the wrong column. The flow across the page should be from the basic statistics (age, sex, etc.), through the social (marital status, education, religion, etc.), to the economic (occupation, work status, etc.). Fertility questions to adult women are traditionally shown on the right-hand side. Enumerators should be told how to proceed down columns and across rows. It may be desirable to complete the first one or two columns first, and then to take each row separately. The best procedure for each questionnaire should be devised and taught.

The essential feature of such a questionnaire is that the number of questions should be limited and that they should relate to simple matters well within the capability of most respondents to answer. Inexperienced enumerators using the one-sheet household questionnaire will be fully extended in defining the household correctly and obtaining basic particulars for each household member. Sensitive questions requiring a skilled probe (which it will not be possible to include on the form) relating to events over a number of years may be beyond them. Many examples indicate the gravity of the problem that may arise. Tape-recordings of actual interviews show that in a substantial proportion of cases the right question is either never asked, or is put wrongly as a result of mistranslation; and that in other cases the right answer, although given, is misrecorded. Thus, 'In about one-third of the tape-recorded interviews concerning eligible women, the question on the date of the most recent live birth was never asked.'[7]

It is easy to put the blame for this state of affairs on the respondent or the enumerator or both. Much of it should properly lie with the surveyor who included the questions requiring highly skilled interviewing. Thus a live birth probe requiring a question such as, 'Did the child cry after it was born?' with a note that if not, it should be recorded as a still-birth, is not suitable when the main input in terms of construction of the probe and its translation into the required dialect is left to an inexperienced enumerator.

The demographer may consider himself caught in a trap—estimation of birth and death rates from the most recent events require very large samples, and thus require a large number of enumerators, many of whom will lack experience. The use of indirect questions (see Chapter 3) or questions covering events over a longer period, together with specially developed analytical techniques, reduces the need for such large samples. A question on orphanhood, for example, is not detecting a rare event; instead it classifies the population according to an attribute that contains large numbers in both the 'yes' and 'no' categories. With a smaller sample, the demographer may use higher grade enumerators and a different format of questionnaire—one discussed in the next section. Finally, this format of questionnaire is inappropriate for most economic questions. A column headed 'Total Income' or 'Regular Expenditure' will not do. If income is to be assessed it must be itemized using a check-list type of questionnaire (see Section 6.7).

6.6 THE VERBATIM QUESTIONNAIRE

The main reason for a rigidly structured questionnaire, in which each question is printed in full in the form in which the enumerator is instructed to deliver it, is that it is seen as vital that the answer given should be to a very precise question, not to an approximation of the question. Often, the question includes words of a technical nature; for example, the World Fertility Survey Core Questionnaire contains many words dealing with intercourse, contraception, and pregnancy that are carefully chosen to have precise meanings. The difficulty is to preserve this precision when communicating with the respondent, who may not be used to the stylized sentence construction and may never have heard some of the words used. Moreover, if the questions are to be put in a precise manner, without changes of wording or emphasis, the questionnaire must be translated into the language of the interview (if this differs from the language used for drafting). To leave the enumerator to translate as best he can at the time of the interview defeats the purpose of the original careful drafting, for slight changes of meaning will inevitably be introduced.

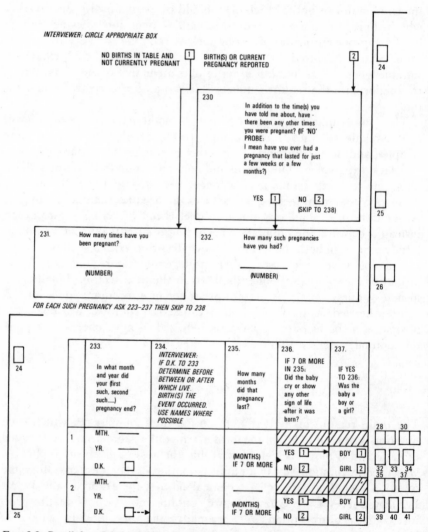

FIG. 6.2 *Detail from World Fertility Survey Core Questionnaire*

The translations into the local languages must be tested. Great difficulty will be experienced in translating certain questions into dialects that do not have the vocabulary to handle shades of meanings that occur in the original. One useful method is to employ one person to translate from the original into the dialect and then reverse the procedure by employing another person to translate back into the original language. If the retranslation does not match the original, further attempts must be made to achieve a better translation. If translation is very difficult, owing to the limitations of the local language, the surveyor should take warning that the respondents may not understand the questions. The survey may be assuming an aspect of linguistic sophistication inappropriate in the circumstances. If the respondent is puzzled by the question, but the enumerator is constrained in the clarification he can offer, puzzlement may turn to bewilderment and annoyance. On the other hand, to allow the enumerator freedom of explanation defeats the purpose of the fixed wording on the questionnaire, introducing, as it does, uncertainty about how the enumerator finally phrased the question. Another problem is that a variation of spoken emphasis, with no change of words, may alter reactions.

A verbatim questionnaire must take account of the different channels a probe may follow according to the variation in response at each step. This leads to the 'skip' instruction and the branched question. An example is shown in Figure 6.2. It will be seen that even the probing questions are spelt out—the enumerator is acting as a mouthpiece and is not reacting spontaneously to the reply. Note, also, that this type of questionnaire may revert on occasion to the one row per event, with headed columns.

This format certainly makes for a lengthy questionnaire—in terms of the number of printed pages, if not in time. The layout needs very careful draughtsmanship if it is not to take on the appearance of a maze-like mess. The example also illustrates how easily the surveyor can become an inquisitor. The enumerator sticking to the 'script' may, even though carefully trained, turn questions such as 236 into an upsetting experience for the respondent.

The skip instruction should be kept simple and, unless it is dependent on the answer to the previous question, it becomes difficult to administer. Its usefulness is very doubtful in the following example taken from a recent survey:

2.3.1 Do you have a job or business or do you normally work on the farm belonging to the household, but did not work yesterday?

YES/NO ⟶ GO TO 2.4.1 IF 2.2.1 EQ NO
GO TO 2.3.5 IF 2.2.1 EQ YES

And, later:

Were you seeking work (employment) yesterday?

YES/NO

$\downarrow$

GO TO 2.5 IF 2.1.1, 2.2.1, & 2.3.1 EQ NO

This need to check previous answers before knowing which way to 'skip' is not to be recommended. Note also the convoluted form of the question 2.3.1. Will the respondent grasp the complete question, or will he only retain the culminating phrase and answer that?

If skip instructions are to be used, the designer needs to draw a flow chart to ensure that all possibilities have been covered sensibly. In the following example there are several ambiguities; but, in any case, do we need to ask a young man who has been in his first job since school for two months if he looked for work last week? The reader may care to trace other possible combinations through this sequence, which in the original was in the form of columns, making the process even harder.

42. Did you work or have a job last week? YES/NO If 'yes' go to 43, if 'no' go to 44–45.

43. Have you been working in your present job 3 months?

44–45. When last did you work (In completed weeks, months, or years)?

46. If 'yes' in Col. 43 or 3 months in 44–45 is/was this your first job? If 'yes' go to 48, if 'no' go to 47.

47. Why did you leave your last job?

48. Did you look for work last week? YES/NO

49. If 'yes' in Col. 48 how did you look for work?

50–51. If you neither worked nor looked for work last week, when last did you look for work?

52. Did you look for work for the first time within past 6 months?

53. If you neither worked nor looked for work last week why did you not look for work?

By this time the respondent and most enumerators will be completely befuddled! Note also that the reply in columns 44–45 can be in any one of three units with no instruction on the form itself to specify which has been used.

The following examples indicate how far the format may be taken and the dangers of trying to exploit the interview situation as much as possible.

Example: How many important people in this country do you think are using any of these methods to delay or prevent a pregnancy? Would you say many, some, or none?

Could much use be made of the data from this question?

Example: What kind of work did your husband's father do, when your husband was about 12 years old?

This probe relies on a weak and extended link.
In the absence of the person concerned the following question is, in our opinion, undesirable (and will not, in any case, yield reliable data).

Example: Could you tell me what was your husband's average monthly income during the last twelve months?

Other forms of the verbatim questionnaire do not use skip instructions and rely on a series of questions in logical order with no branching instructions at all. The layout in such cases can be made more pleasing both to the eye and pen. The desired effect is that each question should be printed on one line commencing on the left-hand side of the questionnaire and extending no further than one-half to two-thirds along the width of the sheet. The 'dotted line' or box for recording the response comes next, and, if pre-coding is not possible, the right-hand edge should be reserved for the code boxes to be used by the clerical staff in the office. Two or more questions should not be placed on the same line. Responses recorded at varied points across the sheet cause confusion and omissions, both in the field and in the office. The erratic dotted line—where the space for the answer on the right-hand side is linked to the wrong line on the left-hand side of the sheet—should be avoided.

It is not always easy to achieve a simple layout, but much can be done with thought and experience. The commonest cause of an untidy layout is that some questions are much longer than others. If the enumerator is allowed discretion in the manner of posing the questions, it is usually possible to achieve the desired symmetry. This cannot be done if the precise wording of the question is important, particularly if some of the vernacular translations are much longer than the original. This is a drawback that must be considered before deciding on a rigidly structured interview.

Whatever the need for symmetry, the space on the dotted line for the answer must be large enough. It is no good expecting the enumerator to record the crop name in a space such as □. This may be sufficient to record 'Tea', but what about 'Finger Millet'? This is a similar error to that of inappropriate column widths in the household questionnaire.

An excerpt from a Zambian survey[8] shows the regular layout that should be aimed at:

10. Maintenance and repair of
    (a) Plant and machinery   K . . . . . .
    (b) Transport equipment   K . . . . . .
    (c) Farm buildings and structures   K . . . . . .
    (d) Other constructions (except land improvements)   K . . . . . .

11. Purchases and repairs of hand tools and small
    equipment used during the year of small
    inexpensive value   K . . . . . .

12. Insurance, taxes, rent, interest   K . . . . .

13. Other expenses (post and telegraph,
    telephone licenses, etc.)   K . . . . . .

One might have looked for a shorter form of Q.11, but otherwise the appearance is pleasing. It is notable that this is from a questionnaire sent by mail to large farmers. We stress once more that as much care should be given to the layout of forms for the enumerator as to those for the general public.

6.7 THE TABULAR QUESTIONNAIRE

This type of questionnaire may contain no actual questions at all: it is in the form of a two- or three-way table to be filled in by the enumerator. It is the type most commonly used in agricultural surveys. The need for the enumerator to cover the whole range of crops grown, livestock raised, and farm inputs and outputs, explains the popularity of this type of questionnaire, which serves as both an *aide-mémoire* and a recording sheet for the enumerator, without detailing the wording of the questions. The enumer-

Costs during the last 30 days (since last visit)

Purchases of animals (specify no. & type)	Cash payments	In kind payments (specify goods and quantity)
Bone meal		
Water fees		
Grazing fees		
Veterinary fees & requisites		
Other (specify)		

FIG. 6.3

ator is allowed wide discretion in deciding how to word the question and the choice of explanations and probes that may be required. From this point of view it is the antithesis of the type of questionnaire considered in the previous section.

Surveyors may feel that they are abrogating their responsibilities if they give too much freedom to the enumerator. Its suitability depends on the topic, the nature of the respondent, and the training of the enumerator. It should certainly not be used unless the enumerator can be rigorously trained. The extent of the discretion granted to the enumerator should be carefully delimited during the training; and close supervision is required to see that these limits are adhered to.

If the questions relate to numeric information, the dangers inherent in this procedure may not be too great. The numbers of livestock by age group, sex, and type present problems in collection, but a well-trained enumerator may be expected to pose questions without distortion. On the other hand, information in income and expenditure, although numeric, presents grave definitional problems that can only be accommodated within a tabular questionnaire by careful standardization of approach. If income is to be assessed it must be itemized, so that the enumerator has a 'check-list' of components of income to be probed, explained,' and recorded. Total income must be obtained by a process of addition, not by an inspired feat of intuition and understanding between enumerator and respondent.

The layout shown in Figure 6.3 is not satisfactory. It has an air of being an early draft: units are not specified; the number of animals purchased is in the same column as the type; and the types are not specified, although other items of expenditure are. Figure 6.4 shows an example of a layout that experienced enumerators have found easy to use.

The other common form of questionnaire that fits within this category is the listing of items for a consumption or expenditure survey. Figure 6.5 shows a segment of one of this type.

The routine listing of items provides little scope for the imagination of the draughtsman; his skill in this instance has to be exercised in blocking the items so that the tedium of a long simple list is broken by natural groupings demarcated by heavier lines, or some similar device. The specification of the items to be recorded has to be considered. One or two lines must be left for 'Other, not elsewhere specified'. The list should be such that these lines are only used for the genuinely unexpected. In one recent survey of expenditure, tables (two to a page) were provided for each general type of expenditure. Very few items were specified separately, and many empty rows were left for 'other' items. Indications of what these might be were given as a footnote. Thus, for example, under the heading 'Recreation and Amusement', cinema, theatre, toys, radio, and

KENYA GOVERNMENT: CENTRAL BUREAU OF STATISTICS: RURAL SURVEY 1976-77.

confidential.

SALES OF AGRICULTURAL PRODUCE: FORM M3

PROVINCE _____ DISTRICT _____

LOCATION _____ DATE _____

CODE [1 | 2 | 3 | 4 | 5 | 6 | 7 | 8] CARD [9 | 10] CYCLE [11 | 12]

SALES OF AGRICULTURAL PRODUCE FOR THE PAST 7 DAYS. From:................ to:................

DATE	ITEM	UNIT OF SALE	NUMBER OF UNITS	VALUE OF SALE	WAS PAYMENT MADE IMMEDIATELY?	TO WHOM WAS ITEM SOLD?	WHERE WAS POINT OF SALE?	WHAT TYPE OF TRANSPORT WAS USED?	DID SELLER PAY FOR TRANSPORT?	IF YES, WHAT WAS COST OF TRANSP.?	DISTANCE TO POINT OF SALE (km.)	TIME TAKEN TO REACH POINT OF SALE (hrs)	NAME AND ADDRESS OF CUSTOMER (IF NOT KNOWN GIVE THE NAME OF THE POINT OF SALE).
13	14	15	16	17	18	19	20 21	22 23	24 25	26	27	28	29	30	31	32	33	34 35	36 37	38	39	40	41	42	

CENTRAL BUREAU OF STATISTICS - MINISTRY OF FINANCE & PLANNING
URBAN FOOD PURCHASING SURVEY 1977
FOOD EXPENDITURES FOR HOME CONSUMPTION: FPS 5   Confidential

Cluster No.　　Household　　Card　　Visit　　City / Town............ Name of Cluster....................

　　　　　　　　　　　　　　E　　　　　Enumerator........... Date.................

	ITEM	Amount Shs.	Cts	Cash or Credit	Usual Retailer	Type of Retailer
MEAT	Beef	0	0			
	Goat	0	1			
	Sheep	0	2			
	Pork	0	3			
	Poultry	0	4			
	Tinned	0	5			
	Other	0	6			
FISH	Fresh	0	9			
	Dried	1	0			
	Tinned	1	1			
EGGS	Number	1	4			
Milk	Fresh Packeted	1	7			
	Fresh Unpacketed	1	8			
	Dried	1	9			
	Tinned	2	0			
	Cream / Cheese	2	1			
OILS & FATS	Butter / Margarine	2	4			
	Cooking	2	5			
	Other	2	6			
SUGAR	Refined	2	9			
	Cane	3	0			
	Jams	3	1			
	Sweets	3	2			
BREAD & CAKES	Bread	3	5			
	Biscuits	3	6			
	Cakes	3	7			
CEREALS	Green Maize	4	0			
	(Maize (Grain)	4	1			
	Maize Meal	4	2			
	Flour	4	3			
	Rice	4	4			
	Sorghum	4	5			
	Millet	4	6			
	Packeted	4	7			
	Other	4	8			
SPICES	Salt / Pepper	5	1			
	Flavourings	5	2			
	Exotic	5	3			

	ITEM	Amount Shs.	Cts	Cash or Credit	Usual Retailer	Type of Retailer
BEANS & PULSES	Beans (Green)	5	6			
	Beans	5	7			
	Peas	5	8			
	Other	5	9			
VEGETABLES	Cabbage	6	2			
	Sukuma Wiki	6	3			
	Other Leaf Veg	6	4			
	Tinned Veg	6	5			
	Carrots	6	6			
	Onion	6	7			
	Other	6	8			
FRUITS	Tomatoes	7	1			
	Oranges	7	2			
	Mangoes	7	3			
	Bananas (Sweet)	7	4			
	Bananas (Cooking)	7	5			
	Pineapple	7	6			
	Other	7	7			
ROOTS	Nuts	8	0			
	English Potatoes	8	3			
	Sweet Potatoes	8	4			
	Cassava	8	5			
	Other	8	6			
BEVERAGES	Coffee	8	9			
	Tea	9	0			
	Sodas	9	1			
	Beer	9	2			
	Spirits	9	3			
	Other	9	4			
OTHER PACKETED & PROCESSED	Soups	9	5			
	Chips etc	9	6			
	Exotic / Import	9	7			
	Other	9	8			
TOTAL OTHER		9	9			
OTHER (SPECIFY)						

FIG. 6.5 *Source*: as in Fig. 6.1

radiogram are listed as items on the first five rows. There follow twenty blank rows. Under the table, nineteen other items are listed as an enumerator *aide-mémoire*, including records, club fees, sports goods, etc. The enumerator is less likely to miss expenditure on these items if they are listed in the table.

6.8 THE ATTITUDINAL QUESTIONNAIRE

In Section 6.3 we gave an example of an open-ended question, 'What constrains you from expanding your production?' and commented on the enumerator–respondent bias that resulted. The danger that this type of bias becomes a serious problem is high if open-ended or attitudinal questions are used.

It is possible to ask a respondent his attitude to a phenomenon that is within his experience, which indeed he may be currently experiencing. In such cases pre-testing should lead to a list of options which can be pre-coded. An alternative is to take the respondent through a set of questions that gradually elicits his opinion or attitude to the topic under study. For example, the following sequence probes attitudes and opinions of different levels of firmness:[9]

519. Do you want to have another child sometime?
520. Would you prefer your next child to be a boy or a girl?
521. How many more children do you want to have?
522. (Currently contracepting or not, from question 505)
523. Have you or your husband used a method to keep you from getting pregnant since the time of your (last) child's birth?
524. What was the last method you used?
525. Did you stop because you wanted to become pregnant?

This combination of questions into current practice, reasons, opinions, and attitudes may elicit reasonable data. Each question is capable of being answered in a simple 'closed' fashion.

It is almost impossible to produce a good layout of the questionnaire if the questions are open ended. Should the space for the answer extend to one dotted line or two or more? This type of question may be useful at the pilot stage, or in a case study where the surveyor can read, digest, and condense the answers of all respondents. They should be excluded in surveys of large numbers of respondents except for a 'safety-valve' question that allows the respondent to air his views.

Questions that demand a judgement by the respondent that he is not in a position to make should not be put. If a remotely situated, small, mainly subsistence farming community has had little experience of official agricultural advice there seems little point in asking members of it for opinions about detailed improvements in the extension service of the department of agriculture. Little will be learned, except perhaps the opinion of the enumerator! In a similar category may be placed hypothetical questions of the type, 'If you were to change your job, what would you . . . ?' The World Fertility Survey adopts this type of question in the following examples:

Do you think you and your husband may use any method at any time in the future so that you will not become pregnant? [Asked of women who have never used a contraceptive method.]
If you could choose exactly the number of children to have in your whole life, how many children would that be?

It is not easy to transpose oneself into a hypothetical situation and give a considered reply; and any reply may be an unreliable guide to what may actually be done if the situation does arise. Moreover, the hypothetical situations envisaged are likely to vary from one respondent to another.

Even if a range of options is provided in the form of word-pictures, this type of question may well misfire. Some of the possible answers conceived beforehand by the surveyor may in the interview appear patronizing and give the respondent, who is usually quick to sense this, a jaundiced view of the survey. An example is a question that sought the attitude of respondents to a new road built to their village. The scale of answers was intended to range from very negative to very positive. The options started with the extreme negative reaction given as 'The new road to your village is:
(a) A snake bringing death and unhappiness to the area. . . .' We conjure up the happy thought of a respondent replying 'No. But the cost–benefit advantage is marginal'! Similarly, qualitative scales such as 'good', 'average', 'poor', or 'better/same/worse than last year' frequently provide only weak data unless they are carefully controlled. One man's mean is another man's median.

6.9 DESIGN FOR ANALYSIS

A questionnaire design that is convenient for the enumerator and respondent may also be convenient for processing the data; but the demands of the editor, coder, and data entry clerk usually impose additional requirements on the designer.

It is common to number the questions (see examples quoted earlier), both as an aid to communication between office and field and as a reference at the processing stage. A frequently used system is to give each section of the questionnaire a number and to number the individual questions as 'decimals' of the section number. For example, section 3 on farm inputs will start with question 3.1 and end, perhaps, at 3.14.

Enumerator entries should, where possible, be in digital boxes to facilitate data entry on to a computer file. Other entries should provide for office coding on the questionnaire itself. If the data have to be transposed on to a coding sheet, not only is there a cost in time, but an additional possibility of error is introduced—the copying error. Basic data management software available on microcomputers now allows for data entry to be undertaken using a facsimile of the questionnaire on the screen—the data being typed into the computer in a format identical to that on the questionnaire itself. The need for an uncluttered and well-structured questionnaire design thus becomes doubly important.

The use of a different colour or shading for the boxes containing data to be punched is recommended. Beware of shading that is so dark that it obscures the entries made on it! No one would plan for this; but the intention of the shading should be made clear to the printer.

Notes

1. Kendall, M. G., and Buckland, W. R., *A Dictionary of Statistical Terms*, Oliver & Boyd, Edinburgh, 1960.
2. World Fertility Survey, *Core Questionnaire and Modifications*, Basic Documentation Nos. 1 and 10, London, 1975 and 1977.
3. *Enquête permanent auprès des Menages, République de Côte d'Ivoire*, Ministère de l'Économie et des Finances, 1985.
4. Raj, D., *The Design of Sample Surveys*, McGraw-Hill, New York, 1972, p. 216.
5. Zarkovic, S. S., *The Quality of Statistical Data*, FAO, Rome, 1966.
6. Blacker, J., and Brass, W., 'Experience of Retrospective Demographic Enquiries to Determine Vital Rates' in Moss, L., and Goldstein, H. (eds.), *The Recall Method in Social Surveys*, Institute of Education, University of London, 1979.
7. Ibid., p. 57.
8. Agricultural and Pastoral Production, Republic of Zambia, 1977–8.
9. World Fertility Survey, op. cit.

7

The Team

'In your own case' said I, 'from all that you have told me it seems
obvious that your faculty of observation is due to your own
systematic training?'

The Greek Interpreter

7.1 THE TEAM SPIRIT

The quality of the data collected during a survey depends above all on the
quality of the field work; and this, in turn, depends on the ability of the
surveyor to create and sustain morale in the field. This cannot be done by
sitting in a central office and issuing exhortations, instructions, and
reprimands. Specialist inputs, even those in the statistical sphere, such as
sample design and data processing, can be partly contracted for. The
execution of the survey in the field does not necessarily require high
qualifications, but it requires someone who can develop a proper rapport
with people—particularly the enumerators and respondents—and so
build and maintain an effective field organization. The appropriate
attitude of mind has already been indicated in our discussion of the
questionnaire, where we have emphasized the continuous and close
attention that must be given to the needs of both enumerators and
respondents. This does not mean slack control and an acceptance of the
second best: it means the setting of standards that can be maintained,
combined with encouragement and support of the enumerator.

The authority of the main surveyor is often required on the spot to
rearrange timetables, to reallocate staff and transport, and to solve any
difficulties with local officials and with respondents that threaten the
survey. The knowledge that the surveyor is accessible, that he will respond
readily to requests for help and a visit, and is in any case likely to be
calling, will stimulate effort and a willingness to work in the same spirit.
Although it is not well documented, everyone with experience of survey
work knows that apparently similar surveys with similar questionnaires
covering much the same topics can have very dissimilar results because of
the different attitudes taken by surveyors to their role in relation to the
field work. A trip to an isolated enumerator is not a waste of time, even if
the journey is long and everything is found to be in order. The good
surveyor is like the good commander: he has the qualities that make his

team want to succeed in the face of difficulty. The surveyor who retires to head office at the end of the training courses and waits for data to flow in will not run successful surveys. The surveyor who is not himself involved in the training courses is not worth talking about.

The difficulties of control will increase as the size of the operation extends. The case study and the small survey will involve only the research worker and one or two assistants, who may have technical qualifications or experience in the topic under investigation. It is relatively easy to manage. At the other extreme, it is almost impossible to generate a single animating impulse to the field staff in a census when thousands of part-time or casual enumerators are being used. The most effective integrating force here will be the sense of national identity, mentioned in Chapter 3.

Calculations of the size of the team required, and the level of transport and equipment it needs, are an integral part of the survey design; and, as indicated in Chapter 2, they have to be discussed and agreed upon in the user–surveyor dialogue. Recruitment and ordering programmes must be put in hand to ensure that the agreed requirements are met. If the approved budget is suddenly reduced, or rapid inflation alters the situation, the scope and design of the survey must be reviewed and, if necessary, adapted or completely rethought. The levels of manpower and equipment (both in quantity and quality) originally allocated will almost certainly have been reduced to the minimum during the budget discussion; further reductions will result in the failure of the survey, unless the cuts are offset by appropriate changes.

7.2 THE ENUMERATOR

The selection of enumerators requires careful organization. The selection procedure should consist of a test and an interview. The test can replace or supplement educational prerequisites, and may be used as a screening device if the advertisement of enumerator vacancies brings forward a large number of applicants. Personal characteristics are of primary importance, however, and an interview is needed. Although it is not an infallible guide, an interview provides better grounds than a written test for assessing potential. Further weeding out of unsuitable candidates will be made during and at the end of the training. We recommend that no one should be recruited to a permanent enumerator force until he has proved his worth in at least one full survey.

The attributes required include:
(a) knowledge of the locality (a local man is preferable to an 'import', other things being equal);
(b) fluency in the local language(s);

(c) attractive personality (strangers at the door will not be well received if they are crude or uncouth in their appearance and approach);

(d) adequate education (this may be little more than literacy, basic numeracy, and the ability to absorb training);

(e) ability to work alone (ability to handle the loneliness of the remotely based enumerator);

(f) tolerance, even liking for uncomfortable 'field' conditions;

(g) honesty: dishonest enumerators who will fake difficulties and ease their burden by inventing data occur only too commonly.

It will be a fortunate surveyor who is able to recruit enumerators all of whom possess all these qualities; but many surveyors do not even attempt to do so, contenting themselves with selecting candidates with a secondary school leaving certificate of one kind or another.

The advantages of a 'local' enumerator are obvious, but in some circumstances too local an enumerator may be at a disadvantage. If the topic of the survey is of a personally sensitive nature, the respondent may be more embarrassed talking to someone from the same village than to a 'stranger'. In these circumstances, the criterion that the enumerator should be a local man might best be interpreted as meaning he should come from the same district as that in which he is to work and should be a member of the tribe or clan to which his respondents belong. Fluency in the local language will normally follow if this criterion of locality is observed; it should be checked, for it is an essential requirement. This is self-evident, too; but we have known cases where enumerators were posted to areas where they could not speak the language used by the respondents.

The personality of the enumerator is often neglected in the selection process, yet the enumerator must establish a rapport with the respondent if the interview is to be successful. A man of slovenly personal habits, a shy man, a man of unpleasing appearance, or an aggressive man, will not make a successful enumerator. In a recent survey the selection process resulted in an enumerator post being offered to a man with a severe stutter. Because his education qualifications were good it had not occurred to the officer in charge that it would be impossible for this man to conduct an interview. Remember, also, that in many household surveys the enumerator will be *resident* in the area for some time; if he acquires a bad reputation in the area owing to his unofficial activities, he will soon find himself cold-shouldered by the respondents.

The sex of the enumerator is not of major relevance as a selection criterion so far as abstract efficiency is concerned. Male enumerators have handled fertility questionnaires addressed to female respondents success-fully, and female enumerators have travelled from farm to farm interviewing farmers on their production and sales of crops. It is generally advisable to use female enumerators for surveys dealing primarily with women in the population, especially for fertility and food consumption surveys. Among certain Moslem populations the use of women to interview women will be a necessity. There are, however, circumstances where a surveyor may hesitate to use females. A requirement to live and work in certain areas may be difficult or dangerous for the unaccompanied woman. A young female enumerator making evening calls on respondents in the less salubrious streets of a city may be facing higher risks than a male enumerator. Our experience is that when female enumerators can be employed, they are often more diligent, careful, and accurate than their male colleagues; and we think the tendency to discriminate against women as enumerators, even when circumstances are suitable for them, uncalled for and ill-advised.

The required education level for enumerators is a much disputed topic. Enumerators engaged in land measurement must be able to use a compass and/or range-finder or similar device. Other surveys have other needs; clearly, the minimum educational level required rises as the complexity and technical content of the survey increase. It helps, too, if the enumerator's education and experience give him some knowledge or understanding of the subject of the survey. For example, if the survey is of costs of agricultural production, or the use of sprays or fertilizers, an enumerator raised on a farm may be preferable to an urban-bred enumerator who has no familiarity with farming procedures. Our general impression is, however, that the required educational standard of enumerators is set too high, and this may be positively damaging. The best enumerators with whom we have worked had received limited formal education; conversely some of the worst have been those with university entry-level qualifications. Over-qualified enumerators tend to get bored with the repetitive nature of the work and are then clever enough to produce 'faked' data that pass ordinary editing checks. They often behave arrogantly towards respondents, or 'lead' them. They tend to resent living in remote rural areas with limited facilities, and they are liable to resign or 'disappear' without notice in the middle of a survey. They are also more likely to find better job opportunities. This is not to say that there are no good enumerators with academic qualifications, but a candidate who possesses the other qualities we have mentioned, lacking only a secondary school education, should not necessarily be rejected. Unfortunately, if selection may lead to a permanent post in government or a large institution, the general educational levels for entry may make it impossible to recruit some of the best potential enumerators.

These comments must be interpreted against the specific needs of the survey: the level of the enumerator must match the level of the respondent. If a survey involves sophisticated respondents, for example, personnel managers in a manpower survey, the enumerator must have sufficient education and experience to sustain the interview. An enumerator who works excellently interviewing farmers in their fields may find it difficult to carry out successfully a rent survey in urban high-income districts.

An enumerator often has to live alone, in difficult conditions, sometimes encountering initial hostility from the group he is hoping to enumerate. He is usually paid rather poorly. We have known many who enjoy the life and ask for little more; but it is not surprising that some cannot bear it for a long period. The qualities that enable an enumerator to sustain this life, and those of honesty and diligence, are difficult to assess at a selection interview. Questions can be directed to explore these areas, but inevitably some of those selected will be found unsuitable. More enumerators than are required should therefore be selected at the first stage—perhaps 10 to 20 per cent more—depending on the size of the force required. The training course should help to sort out unsuitable candidates, though the process cannot be expected to be completely successful. A determined attempt to assess potential must be made, since a large proportion of field errors can usually be traced to a relatively small number of bad enumerators. Hastily recruited replacements, with only partial training, will not repair the damage.

This discussion has been on the basis of recruiting an enumerator team specifically for each survey. This was, indeed, the almost universal practice in developing countries until recently. The building up of survey capability provides opportunities for recruiting a full-time permanent enumerator force with security of tenure, and this is already occurring in some countries. There are advantages and disadvantages. The big advantage is that good enumerators are retained; the training and experience gained in one survey are not lost but can be carried forward to the next. Much of the skill of an expert enumerator is his ability to establish a rapport with the respondent. This skill grows with experience, and it is clearly desirable to retain those who have developed it. The disadvantage is that some poor enumerators may slip into the permanent cadre, or that the good enumerator, once given the security of permanent status, becomes indolent. The nature of bureaucracy is such that once a man is permanent it becomes very difficult to dispense with his services. Moreover, as already mentioned, enumerators who are highly satisfactory for one type of survey may be unsuitable for others. On balance we believe that a small permanent enumerator force is to be desired, but the conditions of entry into this force must be well designed and rigidly adhered to. One condition, already mentioned, should be the successful completion of a probationary period as an enumerator. A career structure, with opportunities to move on or out, is needed to keep the system going over the years. Such a 'core' force should be suppplemented with temporary enumerators recruited on a survey specific basis.

Many surveys are designed to enable enumeration by mobile enumerators, who conduct the interviews in one area as a team before moving on to the next. This method makes supervision easier and enables checks of enumerator biases to be made during the enumeration, without waiting for post-enumeration checks. It may cause difficulties if the team has to 'sweep through' a large part of the country. Language and familiarity problems may occur: it may not be possible to keep the team 'local', as discussed above, in all areas it is covering. Transport and accommodation may also present difficulties. But a limited number of mobile teams is, in general, preferable to a large force of enumerators, static within a cluster. Morale is better, peer pressure improves performance, and greater sampling efficiency can be achieved. On the other hand, the need to cover a substantial number of clusters (see Chapter 4) means that even if enumerators work alone, some mobility is necessary—it may be useful to think in terms of motor cycles or bicycles for the enumerators, depending on the terrain to be covered and the existence, or otherwise, of public transport.

Limited mobility is appropriate for surveys requiring more than one visit to each respondent over the year, in order to monitor change which is due to seasonal and other factors (see multi-visit surveys, Chapter 4). In these surveys there is great advantage in having the enumerators comparatively static, handling a few enumeration areas within the same district. This procedure helps to ensure that the enumerator is a familiar figure in the area, and, if he is a good enumerator, will lead to a close relationship between him and his respondents which will stand the survey in good stead.

A detailed schedule of work should be provided. Enumerators are often either underemployed or overemployed; both conditions are detrimental to the survey. The main reason for suboptimal deployment of enumerators is a failure on the part of the surveyor to foresee the way in which the work-load will vary according to the conditions prevailing in the enumeration area. Five respondent visits a day may be a proper work-load in one enumeration area, but one or two per day may be all that can reasonably be expected in a larger area, containing the same number of households, but with few, if any, roads or tracks. To set an average of three respondents per day for each enumerator results in idle time, leading to boredom, for one, and stress and undue strain for the other. Potential difficulties of this kind need to be assessed before adopting sample designs which generate equal loads per enumerator.

Scheduling of work is another argument in favour of keeping the permanent enumerator force small. Large numbers require sophisticated administrative and management support and impose a need to generate new surveys at a sufficient pace to keep the field force occupied. Too often the creation of a national survey capability has become synonymous with the conduct of a continuous round of surveys, initiated to keep the survey force going rather than to fulfil national needs.

It is the enumerator's duty to preserve the completed questionnaires (until collected or dispatched) in such a way that they are not damaged by dirt or weather and are secure from unauthorized persons. Enumerators should be given the means to fulfil this responsibility. Clip-boards with plastic protection are useful as work surfaces, and a box with a padlock is necessary for storage. This is a minor but frequently neglected preparation.

7.3 THE SUPERVISOR

The supervisor still seems to be regarded by some surveyors as almost an unnecessary luxury. It is argued that such supervision as is required can be provided by senior staff making occasional field trips. For very small surveys this may be the case, but supervision on a full-time basis is essential for surveys employing several teams or many individual enumerators. Lauriat provides an example of how increased supervision provided an improvement in data quality resulting from 'an additional six supervisors . . . increasing the original 9 to 15, since it had become apparent within the first year of the survey that the number of enumerators that could be effectively handled by one supervisor had been seriously over-estimated.'[1]

The duties of a supervisor are:

(a) acting as the point of contact between the enumerator and his colleagues, in order to counteract the loneliness of the enumerator working in remote areas;

(b) maintaining time schedules and co-ordinating work in a set of areas;

(c) checking the enumerator's interviewing ability by repeating or sitting in on interviews;

(d) editing the completed questionnaires for misunderstandings and recording errors;

(e) visiting respondents who have questions that the enumerator has been unable to answer and visiting community leaders if there is need for further encouragement or trouble-shooting;

(f) acting as a postman and courier—transfer of equipment, documents, and instructions from office to field and vice versa is often best accomplished by hand carrying;

(g) watching and listening for potential trouble and difficulties that need to be resolved quickly;

(h) covering in emergencies for enumerators who fall sick, making temporary reallocations of schedules, and contacting the next higher level for more permanent arrangements to cover staff breakdowns;

(i) assisting in any post-enumeration check.

These duties require experienced, responsible, intelligent men or women.

Previous field experience, preferably as an enumerator, is the essential prerequisite.

The poor supervisor will narrow his duties to (a) and (f). Too often (c), (d), and (e) are neglected, although they form the very heart of the job. Errors in the recorded data detected during the editing of the questionnaires in the analysis office may be impossible to correct; the data are literally irreplaceable. Examples are the areas of plots that have been harvested (and which had no permanent boundaries such as fences or hedges), and the consumption of food based on a short recall period now well past. The detection of errors in the field, while it is not too late to make corrections, is important. Lip service is paid to this in most surveys, but in practice investigation reveals that the supervisor is often little more than a postman, delivering and recovering questionnaires without giving them more than a cursory glance.

If mobile teams of enumerators are used, one supervisor should accompany each team. If enumerators are dispersed around a clustered sample, the number that can effectively be overseen by one supervisor will vary according to the type of survey and the distance between enumeration areas. Experience has led to commonly adopted ratios in the range 1:3 to 1:6.

7.4 OTHER TEAM MEMBERS

If a survey programme is to be undertaken on a nation-wide and repetitive basis it may be necessary to set up regional or provincial offices manned by survey co-ordinators and clerical assistants. Survey management is often neglected in large surveys. Large field forces of supervisors and enumerators require a support system involving personnel and logistic management and the conduct of liaison activities between the data collectors and the professional surveyors. It is not a matter of duplicating the surveyor's survey design and sampling skills at each regional office, but, rather, staffing these offices with experienced field managers. One reason population censuses work well is that the regional census officers are administrators who know how to run things. Statistical offices have been slow to realize the importance of such administrative skills. A good sampling theorist may be a poor manager of a field operation. Good survey administrators are essential to the team.

Until very recently, there was another universal part of the survey team that prepared the data for its transmission to a more general data processing centre—namely the data editors and coders. In many surveys these roles are still performed in the traditional way, namely by eye and hand. However well trained and supervised the enumerators, it must be expected that there will be substantial errors in recording information on questionnaires. The maxim must be to validate and 'clean' the data at as early a stage as possible—preferably in the field, or immediately on receipt in the survey office.

The qualities required of editors are distinctive. There are those who can scan numeric data quickly and spot inconsistencies and there are those who will never do the job well. An editor must have an 'eye' for detail and emerging patterns. It is the difference in enumerators' patterns of data that may reveal confused or dishonest enumerators. For example, in a multi-visit food expenditure survey, the pattern in most households may be for sporadic and variable purchases of certain commodities, but for one set of households covered by one enumerator the pattern is one of regular and consistent purchases. Is the enumerator faking his data? This can now be checked while there is still time to limit the damage, if such proves to be the case.

The other quality required of editors, and even more so of coders, is the ability to concentrate over long hours on a job that is boringly repetitive in nature. If concentration lapses, not only do errors pass through undetected, but others are introduced.

But, as we implied above, change is under way, brought about by the microcomputer and user-friendly data editing and validation software. With these tools the editing and validation of data becomes an integral part of the process of data file creation. The mind-numbing repetitive checks on each data item are carried out automatically, with the operator being alerted only if the data fail the prescribed tests. The use of microcomputers allows the office members of the survey team to be trained to undertake a more complete and satisfying task that combines data editing, validation, file creation, and manipulation. No longer need the data be passed to a separate computer section or bureau—the survey team can see the job through. Already surveys are being conducted where the microcomputer is used by the data collecting team who enter and validate the information in the field.[2]

Because of the increasing integration of the various aspects of data preparation with those of data processing, further discussion of the development is included in Chapter 9.

7.5 THE TRAINING OF ENUMERATORS

It is, of course, necessary to train *all* members of the survey team, although it may be difficult to get some of those in supervisory grades (and above) to accept this need. But since the main training, and the most difficult, is that of enumerators, we shall include all the points we wish to make about training under this head. It will be seen that most of them are relevant to the training of supervisors, editors, etc., even though the context may be slightly different.

The number of training courses and their length need careful consideration. The pace at which training can proceed will depend very much on
the care that has been taken with selection. Training that will not only
impart knowledge but also overcome individual misunderstandings cannot be conducted satisfactorily with too large a number of trainees. A
maximum of 40 to 50 at any one training course should be the aim. If this
means that a series of courses must be held, they should be organized on a
provincial or regional basis. Co-ordination and standardization are
required, but this should not prevent the discussion of local problems.

The required length of time for the training is often underestimated for
a very simple reason. Teaching students to pass an exam might be
considered successful if a high proportion of them achieve a 'credit' mark
of say 60 per cent. Teaching enumerators to conduct a survey during
which they will be working individually requires that every trainee has a
100 per cent understanding of the methodology and content of the
questionnaires. An enumerator who has understood most of the questions
that he is later to put to respondents is no use; he must be able to 'pass an
exam' on the questionnaire with full marks. The training must cover the
concepts, definitions, and structure used in the questionnaire; it must
include the style of interviewing required and give instruction on the way
measurements are to be made. Courses have to be matched to surveys, but
an outline of a training course for a household survey is shown below as an
example. The survey questionnaire contained five forms of varying
complexity.

Day 1:   Introduction; background to survey; duties and
                responsibilities.
Day 2:   Interviewing techniques and measurement
                instructions.
Day 3: a.m.   Instruction on Form 1—individuals and households.
       p.m.   Classroom exercises.
Day 4: a.m.   Instruction on Form 2—earned income.
       p.m.   Classroom exercises.
Day 5: a.m.   Instruction on Form 2 cont.—unearned income.
       p.m.   Classroom exercises.
Day 6: a.m.   Instruction on Form 3—expenditure (non-food).
Day 7: a.m.   Instruction on Form 4—expenditure (food).
       p.m.   Classroom exercises on Forms 3–4.
Day 8: a.m.   Instruction and exercises on Form 5—assets.
       p.m.   Revision of questionnaires with discussion.
Days 9–11:   Field exercises.
Day 12:   Review of exercises and final instructions.

Details will vary, but we believe that the training of 'raw' enumerators for
most detailed household surveys requires about two weeks. Each topic

may be handled in one or two days with extra time for trial interviews in field conditions. If the questionnaire consists of one form, simple to complete, the training course will be shorter, but it is not normally possible to carry out successful training in less than three days.

The enumerator's ability to understand the questions at the first explanation is likely to be less than the surveyor anticipates. He and his immediate assistants have been immersed in the design of the survey for some time. Definitions that now seem second nature to them will not be obvious to an enumerator. Repetition of instructions will be required, with opportunities for questions and the careful working out of examples in a classroom context.

Each trainee should have copies of the questionnaire and an instruction manual. The lectures should follow this manual, which should explain concepts, definitions, and methods in great detail. At this stage the manual is essential; but the surveyor must make sure that by the end of the training course the necessary instructions are in the enumerator's head. The enumerator cannot keep referring to a manual during an actual interview with a member of the public, for any hope of a natural rapport between an enumerator and the respondent would then disappear. The main use of a manual during the survey is as a reference for a list of permitted codes or abbreviated entries.

The teaching of the enumerators must proceed point by point, survey question by survey question. The lecturer must remember that full understanding by all the students is the only acceptable end in view. This may make for slow progress and a repetitive delivery, but that has to be tolerated. Examples of how to enter the data on the form should be shown in detail. Projection slides of each part of the questionnaire for demonstration purposes are very useful; if they are not available or cannot be used, each part of each questionnaire has to be transcribed on the blackboard. Classroom discussion should be encouraged. This will bring out points the surveyor may have overlooked or dealt with too rapidly; and it should reveal those points that have not been understood.

The extent to which interviews can be simulated in a classroom environment varies according to the local circumstances. In general, it is difficult to achieve the correct atmosphere. The latent acting ability of some enumerators may be revealed in their readiness to accept such a simulation, but most will do themselves less than justice. Realistic field exercises are essential; but these also present difficulties. If the training course has brought together enumerators who speak different dialects it may not be possible to find suitable local 'training respondents' speaking some of these dialects. Interviews during field exercises may be awkward for the respondent, too, since the enumerator is, or should be, under direct supervision and subject to possible interruption during the conversation. Moreover, the respondent may be asked to put up with more than one interview on the same topic with different enumerators. It is important, therefore, that these training respondents are properly briefed and their full collaboration obtained. This is a case where financial reward may be justified. If full field exercises cannot be organized at the site of the training course they should be organized at the regional level before the survey commences. There is no substitute. If the survey involves technical measurements, such as taking the height, weight, and mid-arm circumference of young children, then practice in the required techniques is essential. This is usually easier to organize than realistic sample interviews, since the enumerators can measure each other for the initial practice.

Towards the end of the course trainees who cannot reach a satisfactory standard should be identified and removed from the course. A certain ruthlessness may well be required in this 'weeding out', but if the success of the survey is not to be jeopardized it has to be done. If the precaution of commencing the training with numbers in excess of final requirement has not been followed, replacements will have to be trained individually by a supervisor or field co-ordinator. But this is a dangerous situation since these late recruits may turn out to be no better than those they are replacing.

The surveyor should take pains to cultivate a team spirit during the training, notwithstanding the environment of testing. This is his best opportunity to establish standards, create a good morale, and generally construct a team. Time taken during the classes to explain the reasons for the survey and leisure time contacts between the surveyor and his team will be amply rewarded. To this end, it is desirable to hold the training course at a location where all concerned, trainers and trainees, are resident on site. Courses run in a headquarters in a big city, where everyone goes his separate way at the close of each day's training, are not as satisfactory as those held in a smaller remote spot with residential facilities. This kind of location is also likely to ease the problem of arranging the field exercises for rural surveys.

The end of the training course is often a convenient stage at which to issue to successful trainees the forms and equipment they will need, together with any formal identification documents that are thought necessary.

Notes

1. Lauriat, P., 'Field Experience in Estimating Population Growth', *Demography* 4, 1967, p. 235.
2. Ainsworth, M., and Muñoz, J., *The Côte d'Ivoire Living Standards Survey: Design and Implementation*, LSMS Working Paper No. 26, World Bank, 1986.

8

The Collection of the Data

This was an unexpected piece of luck. My data were coming
more quickly than I could reasonably have hoped.

The Musgrave Ritual

8.1 THE PRELIMINARIES

The preparation of the 'ground' is an important feature of a survey. Visits
to the area or areas selected are necessary for the following reasons:
(a) demarcation of the selected areas in the field;
(b) listing of the population units (households, etc.) within the area,
and selection of the sample units;
(c) appreciation of local problems that may affect the progress or conduct
of the survey;
(d) publicity among the population.

Demarcation and listing of the selected areas are covered in Chapter 4
and need not be stressed again here, but a word is necessary on the
possibility of local problems. However carefully the survey has been
designed, particular circumstances in certain areas may require adjustment to the methods, the definitions, or the questionnaire. These local
problems must be resolved before the survey is started, otherwise hurried
and inappropriate decisions may be taken, leading to unforeseen complications. Examples of 'local' difficulties include pastoral areas where the
concept of a land area being farmed is inappropriate, religious or cultural
taboos that necessitate the rewording of certain questions, and the practice
of income sharing (co-operative or communal living) that requires a
redefinition of the sample unit. In extreme cases the entire enumeration
area may be inappropriate for the survey, resulting in sample design
complications.

Publicity can be handled at two levels: the general, broadcast to the
entire population, and the specific, aimed at the survey respondents.
National publicity certainly has a place in the preparation of censuses and
large surveys, but the publicity for most surveys should be concentrated in,
and at, the survey areas. Detailed explanations to individual respondents
can be left to the time of actual enumeration, but a description of the
survey and the reasons for it should be made available in advance.
Meetings with community leaders, a public meeting addressed by one of
the survey organizers, and letters to the selected respondents are probably

the most popular and effective ways of achieving the required foreknowl-
edge among the population of the area, and so their approval.

We have stated earlier that ordinary people in most developing countries
are extremely tolerant of requests for information, and often too courteous
or puzzled to voice their misgivings. The surveyor should not be deceived by
an apparent lack of curiosity; misgivings may be present that will lead to
falsification or concealment of certain information. Adequate advance
publicity which allows these reservations to be aired, and reasurrance
given, is essential. Needless to say, assurances once given must be honoured;
the surveyor should not permit promises to be made that cannot be kept.
The public may not immediately make a distinction between an official
department and an enumerator asking questions about matters falling
within that department's responsibility. The inability of the enumerator to
influence directly a change in attitude or policy must be made clear.

After a period of preparation, the survey starts. It is important for the
morale of all concerned (and, therefore, for the success of the survey) that
this launching of the survey is not a halting, erratic operation. Enumerators who have failed to arrive in the correct enumeration area, equipment
that is still sitting in the head office, and packets of questionnaires that
have been delayed in transit, are faults unfortunately all too familiar to the
experienced surveyor. The final stage of survey preparation must pay
particular attention to logistic details. The right men must be in the right
place, armed with the necessary equipment and with no personal problems regarding finances or accommodation in the area.

8.2   THE INTERVIEWING OF THE RESPONDENT

The respondent's first impression of the enumerator will affect the entire
course of the interview. In general, enumerators are too casual about this
opening gambit. It is an aspect of the training (see Chapter 7) that is
difficult to communicate to trainees. The slick patter of the travelling
salesman is to be avoided; but a personal introduction, an outline of what
the survey is about, and a clear statement of how the enumerator intends
to proceed are required. Nothing shows up the enumerator who lacks a
suitable personality more than his initial exchanges with the respondent.
Again, we repeat that the tolerance and courtesy of the respondent may
appear to allow the enumerator to get away with a brusque, inadequate,
and even a contemptuous approach (such as the enumerator who arrives
at a field in which the farmer is working and summons *him*), but the
evidence will be there in the lack of quality in the data collected. In
extreme cases, the antisocial behaviour of an enumerator may have serious
consequences not only for the survey in progress, but for longer term
public co-operation with data collection agencies generally.

The interview must be conducted with confidence. The enumerator should have mastered the topics of enquiry so that he can cope with any eccentricities of the respondent's style of replying to questions without losing the required order and rhythm of the interview. A question put in a stumbling, bumbling manner will receive an inadequate, incomplete, or irrelevant answer.

Confidence must be linked to efficiency—indeed is conditional upon it. The conversational approach, the handling of the equipment for taking any necessary measurements, the visual observations and their recording, all must be stamped with the unmistakable mark of a practitioner who has mastered his task through experience in a variety of situations. The fumblings of a nervous and inadequately trained enumerator lead to a bemused respondent and a garbled set of misleading data.

The effect required is that of a friendly conversation between enumerator and respondent. The posing of questions and the noting of the replies should have the flow and pattern of a dialogue, not the staccato delivery, punctuated by pauses, of the inquisitorial chamber. The enumerator will find this easier to accomplish if he is allowed a measure of discretion over the order and wording of the questions, but this discretion can be effectively exercised only by the enumerator totally familiar with the questionnaire. An actor should be able by opening night to come in with the correct line even if given the wrong cue. Similarly, the enumerator should be able to interpose the appropriate question without disrupting the flow of the dialogue, even if it is following an unusual course.

This relaxed conversational effect is impossible to achieve if the enumerator keeps his head down and his eyes glued to the questionnaire. Even enumerators who know their way around the questionnaire fall into this habit; and it is an unerring, and unnerving, indication of an unsatisfactory enumerator. The interviewer should refer to the questionnaire only to record the respondent's replies and he should know where to record the reply, without having to trace his finger laboriously down or across the sheet of paper looking for the correct space.

The good enumerator will not fail to pick up information supplied by the respondent in a conversational reply, when this information would have been requested in a later question. A friendly, voluble respondent may give the answers to two or three questions even though only the first has been posed. A question such as 'Have you any children?' may produce a reply such as 'Yes. Two. A boy aged six and just starting school, and a girl born last July.' It may be necessary to probe for greater detail or to check that other children have not been omitted; but the wording of the next questions should indicate that the information offered by the respondent has been noted. Thus the good enumerator would ask for the boy's name and add 'When you say he is six, does that mean that he was six on his last birthday?' It is not unknown for an enumerator hypnotized by the questionnaire from which his eyes have never strayed to request information regarding the boy's sex. Respondents should not be asked to repeat information unless contradictions or ambiguities need to be cleared up.

The conversational effect is more difficult to achieve if the enumerator is expected to follow a defined order of questions with a prescribed wording. Such an interview needs skilful handling if an undesirable air of an interrogation is to be avoided. The dangers of hectoring the respondent is one of the disadvantages of this type of questionnaire. Take, for example, the following exchange:

Q: As you may know, there are various ways that a couple can delay the next pregnancy or avoid pregnancy. Do you know of, or have you heard of, any of these ways or methods?

A: No.

Q: Just to make sure, let me describe some methods to see if you have heard of them. [There follows a description of all standard contraceptive practices: at the end of each a question is asked.]

Q: Have you ever used this method?

A: No.

[At the end of the sequence of descriptions, if all the answers are 'No', comes another question.]

Q: I want to make sure I have the correct information. Have you ever done anything or tried in any way to delay or avoid getting pregnant?

Even though enumerators stick to the wording provided it is their manner that will determine whether such an exchange has or has not crossed the border of acceptability.

The authors of the structured questionnaire do not intend to prevent the enumerator from adopting a friendly and flexible approach. In a section on general conduct during an interview using a structured questionnaire it is stated:

As far as possible, the interview should be a conversation rather than a formal interrogation. . . . Maintain continuity and easy flow of conversation. . . . If the woman gives irrelevant answers or starts to tell her life history, do not stop her abruptly or rudely but listen to what she has to say and then steer her gently back to the original question.[1]

The potential harm done by enumerators who stick too rigidly to instructions is commented on in the same review, citing a dialogue for question 407 by which time the marital status of the respondent had already been established:

In the following excerpt, one can imagine the look of incredulity on the respondent's face at the stupidity of the question:

Do you currently have a husband?

What?
Do you currently have a husband?
What?
Do you currently have a husband?
I see . . . Yes.

In one respect the enumerator should emulate the detective; he should be a keen observer of the respondent and his surroundings, and have a fine ear for the nuances of the respondent's replies. The decision whether to probe further, and the choice of wording of a question, may be affected by what the enumerator learns from the demeanour and tone of the respondent. An extreme case of a failure to connect physical evidence with a spoken reply was an enumerator who, shortly after recording a farmer's statement that he owned no livestock, took a step back and fell over a goat tethered inside the hut. Brushing himself off, the enumerator took his leave without asking about the ownership of that goat or the further score or so outside the door!

In the typology of surveys given in Chapter 4 a distinction was made between single- and multi-visit surveys and between those requiring an interview only and those requiring a combination of interview and measurement. These different approaches also require different techniques in dealing with the respondent. The comments above were in the context of a single-visit survey, although the principles of a friendly, conversational approach apply to any contact with the respondent. The success of a multi-visit survey, however, is totally dependent on the establishment of a genuine bond between the enumerator and respondent. The first visit must be used for this purpose. An explanation of the survey, the commitment required of the respondent, and the recording of basic household particulars are probably all that should be attempted at this time. During subsequent visits the enumerator should continue to build a good relationship; formal question and answer sessions should be avoided. Few respondents will tolerate visits two or three times a week to record details of household expenditure if this requires the suspension of normal activities and the presence of an impersonal enumerator. The enumerator must create the impression that the recording of the data is a joint enterprise and that he is *interested* in the details given to him.

Measurements and counts, whether of land, cattle, or kitchen stocks, involve an even greater intimacy betwen the enumerator and respondent. The enumerator is intruding in a very real sense into personal lives. The appropriate style may be similar to that of the extension or social worker—a government worker carrying out duties that are meant to be helpful to the community—although the enumerator's usefulness in this respect is less easy to put across than that of those who are giving advice directly. On no account must the enumerator imply that he is able to provide or promise assistance, but his goodwill must be abundantly clear.

Privacy for the interview is, at the least, desirable, and is sometimes essential. Obtaining it is often difficult. The report of a fertility survey states:

Only 7 per cent of interviews were conducted wholely, or partly, in the presence of the spouse. . . . A larger proportion (15 per cent) were conducted in the presence of other adults, typically, mothers, mothers-in-law, or sisters. Obviously enumerators found it more difficult to remove female intruders. . . . In a few instances, enumerators, sensing that the persistent presence of a mother-in-law would harm rapport, abandoned the attempt and discreetly discovered a time when the relative would be absent and a successful re-visit could be made.[2]

Clearly, the respondent should be the final arbiter of whether the presence of a third party is acceptable, but the respondent may not be aware at the outset of the survey that intimate personal or economic details will be required. The presence of persons from outside the household is almost invariably detrimental to the interview and the enumerator should do all that is tactfully possible to prevent it. A persistent relative may require the rescheduling of the visit, as indicated in the quotation above.

8.3   THE APPROPRIATE LEVEL OF ENUMERATION

This is a somewhat neglected issue. Two mistakes are common. The first is to assume, without careful thought, that all sample designs have as their ultimate stage the same type of basic respondent—the head of the household, the farmer, the shopkeeper, etc. The second fault, which occurs most commonly in surveys of business and industry, is to assume that more knowledge will be centred in an easily retrievable file than is in fact the case.

For achieving the survey objectives, or some of them, it may not be necessary to interview individual respondents. Many sample designs make use of a cluster as a first stage unit—it may, for example, be a village or a locality. The water supply for this area may come from one river; there may be only one source of artificial light—paraffin; and the only health centre in the subdistrict is some ten miles from the village. Is there any need to ask each individual respondent from where water is carried, what is used for light, and which health centre is used when sick? Yet this is often done. A community questionnaire (see Chapter 5) to be filled in by direct observation and by questioning a senior member of the community may result in perfectly acceptable data, and save considerable time and effort during the individual interviews.

Surveys of business or enterprises may be concerned with statistics relating to a very disaggregated level of the firm's activities, for example, a survey of the skilled manpower employed. It may be considered that it is sufficient to draw a sample from a list of firms and call at the registered office to seek the necessary information. A survey budgeted on this premise may soon be in serious difficulties. The head office may be quite a small, albeit prestigious, affair; records are kept at the various factories, technical offices, and district offices. Moreover, some of the details required concerning an individual or group of individuals employed by the firm may only be obtained centrally by the personnel officer's going through the relevant personal files. If it is a large company he will not tackle this job with much enthusiasm or sense of urgency. It might have been better to design a survey that involved interviewing a subsample of individual employees to obtain this personal information.

Another problem involving the level of enumeration can be illustrated by a survey to establish wholesale and retail margins for certain unprocessed foods sold in open-air markets. In order to establish the wholesale margin is it best to ask the farmer who sold to the wholesaler the price at which he sold, and also to ask the retailer the price at which he bought the commodity from the wholesaler? Or is it satisfactory, and perhaps cheaper, to ask the wholesaler directly to state his buying and selling prices? Or to approach all three? If the last alternative is chosen, care may be needed if the enumerator is seen approaching a retailer after previously speaking to the wholesaler. If the first respondent feels that the enumerator does not trust him, trouble may quickly follow.

8.4 PHASING THE ENUMERATION

The timetable for the survey enumeration must be laid down in advance and adhered to as closely as possible. The minimum number of respondents to be interviewed each day and the deadline for completion of the enumeration can be specified in advance, but the time when a respondent is interviewed will of course depend on his availability. Guidance on these matters can be usefully obtained during a pilot survey.

Surveys requiring a single interview with each respondent do not usually demand that the date and time of the interview are controlled. For example, in certain surveys scheduled to last one week it matters little if a few enumerators complete their interviews in the first three days (as long as they are not rushing each interview and failing to cover the ground adequately). If, however, the questions refer to a very short period of time, the phasing of the enumeration may be important. In a food consumption survey the details regarding food eaten in the previous twenty-four hours may depend upon the day of enumeration, as dietary habits may be linked to days of the week, for example, religious festivals or beer parties on Saturday nights. Therefore, an equal number of respondents should be interviewed each day, including Sunday. Similarly, the use of family labour on the farm will be affected by the day of interview if a one-day recall is used; rest days and market days will reveal markedly different behaviour from the others.

Surveys dealing with topics such as consumption, expenditure, or labour normally involve repeated visits to each respondent in order to measure the effect of time over a longer cycle, monthly for income, seasonal for labour inputs, and so on. The scheduling of these repeat visits needs to be carefully considered in order to avoid visiting a respondent according to a cycle and with the use of a recall period that is also linked to his behaviour with regard to the topic under investigation. If the respondent is interviewed monthly, in order to record details of expenditure in the previous week, the interview should not always be in the first week of the month after he receives his salary, for his expenditure in the week after pay-day may be very untypical. The order in which an enumerator visits respondents within each survey cycle should follow a schedule, set by the surveyor, using some form of random, or staggered, selection if the recall period is a small fraction of the cycle period, but should be maintained and repeated if the recall period is equal to the period of the cycle.

Agricultural surveys must be designed with due regard to the agricultural seasons. The hangover of the historical emphasis on national accounts (see Chapter 1) can still be detected in the number of agricultural surveys that are scheduled over a calendar year, although the agricultural 'year' is very different. The most appropriate period for the measurement of crop areas is when all the crops have been planted but none have been harvested. This is difficult when there are two cropping seasons in a year, but becomes impossible when the climate is such that crops are planted and harvested in an almost continuous cycle according to inclination and need. We shall return to this point in Chapter 13 when discussing the collection of data on crops. It is sufficient to note here that the failure to take account of biases introduced by missing part of the area under the crop (not yet planted, or, alternatively, already harvested at the time of the enumeration) is a major reason why enumeration of crop areas has been singularly unsuccessful in many countries.

A regular schedule of interviews in each enumeration area can be threatened if an enumerator falls sick or is absent from duty for some other reason. It is not practical to keep a stock of reserve enumerators like substitutes on the sideline of a football match; the range of language abilities and area knowledge would require a large pool. The supervisor or a district office clerk must be deployed to carry on the enumeration as a temporary measure until the enumerator returns or a fully trained replacement is provided. The point is that this contingency plan must exist in advance so that it automatically comes into operation if required. In the absence of such a plan, the delay before the cessation of enumeration is reported to a level where *ad hoc* measures can be effected may entail the loss of the enumeration area from the survey. The training of the supervisors should include the procedures for dealing with emergencies of this nature.

8.5 QUALITY CONTROL

Quality cotrol measures in many surveys are applied only at the data processing stage. Necessary though these are, they should be regarded as an addition, not a substitute, for checks at the time of data collection, when there may still be time to correct any systematic faults detected. Some supervision is so slight that it does little more than ensure that the enumerator appears to be working. This weakness can be removed by introducing the following checks:
(a) supervision of interviews;
(b) editing of the completed questionnaires by the field supervisor;
(c) consistency checks built into the questionnaire;
(d) verification of the data using different reference periods;
(e) verification of the data using secondary sources.
The role of the supervisor is discussed in the previous chapter, where it is emphasized that he should be much more than a messenger and a postman. The supervisor is the key figure in the conduct of quality control checks both during and after the enumeration (see next section).

Actual survey interviews should be monitored so that enumerator—respondent biases will be detected and corrected. If the questionnaire has not been adequately pre-tested, unfortunate last-minute changes may have to be made. If errors are due to enumerator faults then he will have to be given on-the-job remedial training. In addition to attending sample interviews the supervisor should carry out a quick visual edit of completed questionnaires. The schedule of work of the supervisor must be as clearly drawn up and precisely defined as that of the enumerator. It should include a list of checks to be carried out during each of his field visits; his signature on questionnaires inspected by him should be the certification that he has done so.

An assessment of interviews may also be made if a sample of them are tape-recorded. This method is not without its defects. Equipment is costly and goes wrong; batteries and cassettes may be unavailable. The introduction of the machine may adversely affect both the enumerator and respondent, and prevent the development of an easy attitude; and the recorders are not easy to use properly in outdoor interviews. The supervisor may nevertheless be able to use a few recordings to pin-point bad enumerators or particular sections of the inquiry causing especial problems. The surveyor, who may well have to wait for typed and translated transcripts, will often be able to use them only for a post-mortem assessment of his survey.

The questionnaire should be designed so that the enumerator is constantly prompted to question the accuracy of the responses he is receiving. The internal consistency of the information should be checked during the editing of the questionnaires, but at least some of these checks should be so clear that the enumerator will spot inconsistency at the time of the interview. This can be achieved by suitable ordering of the questions—placing questions that serve as mutual checks close to each other. Some apparently independent questions will not be treated as such by either the respondent or the enumerator, so that they are ineffectual 'checks'. To ask a child's birth date and then for its age in months will not produce a realistic check. When the date of birth has been recorded, the respondent or the enumerator will calculate the age from this date.

If individuals in a polygamous family are being recorded, it may be advisable to list the children of each wife after recording the details of the wife. If all the wives and then all the children are listed, no one child is identified with its mother. Recording the children immediately after the mother also helps the enumerator to spot errors of reporting since he can pick out the child who appears to be too old, or too young, to be born to a particular woman.

Questions about major and secondary occupations may be followed by questions concerning activities in the last week that will check, for example, the relative importance attached to farming, as opposed to casual labouring. It may be worth while to include the questions on last week's activities even if the survey's objectives do not require more than general occupational data. Many rural householders still think of themselves as primarily farmers, even though much of their income accrues from, and much of their time is devoted to, some other occupation. The reverse may also occur; that is, the respondent may take the farm work for granted and concentrate his replies on other activities, even though these contribute less to his income than farming. A listing of recent activities may serve to warn the enumerator that further probing of occupation is required.

The use of different reference periods for various aspects of the survey may not only help to maintain control over the quality of the data on each aspect, but may enhance the value of the summary information that emerges from the survey. The respondent may be asked to estimate his total sales by crop over a one- or three-month period. Specific details about the sale, such as to whom sold, place of sale, price per unit, and transport used may be restricted to those sales occurring within a shorter period, such as a week. In addition, the prices of these commodities can be collected regularly from a sample of traders at the local market-place. The result is a fuller and more informative picture of the production and marketing performance of the farmer. There is, of course, no point in taking these steps unless provision is made to analyse the resulting data sets in proper relation to each other.

The enumerator may be helped to spot omissions or contradictions by using a check-list. As an example, consider a survey with repeated visits to a farmer in order to record details of his agricultural production. If at the start of the survey details of the crop composition of each field are recorded, the questions regarding the harvest and disposal of crops put to the farmer during each successive visit should be linked to a particular crop in a particular field. A general question such as, 'Have you harvested any crops since my last visit?' may produce a quick and superficial 'No' from the respondent. A series of questions such as 'What about the cassava in the field the other side of the stream? Have you harvested any yet?' may remind the respondent that he has indeed partially harvested this field for his own consumption. In order to put the questions in this manner the enumerator must be supplied with a summary of the crops grown on each plot. He can scarcely be expected to remember the layout of each of his respondent's plots, certainly not during the first months of the survey. This summary may be a copy of the original questionnaire on which the crop composition and area of each field are recorded, or it may be a simple listing kept by the enumerator in his field notebook (a valuable item of the enumerator's equipment, but often not provided).

8.6 THE POST-ENUMERATION SURVEY

Any census and many large-scale surveys should include provision for a post-enumeration survey (PES). It is now accepted as an integral part of census methodology. It is generally agreed that coverage and content errors, committed by the large number of inexperienced and quickly trained enumerators used in a census, can be partly assessed by comparisons with a small sample re-enumeration carried out by more trained and experienced staff. Whether a post-enumeration check is desirable for other surveys depends on their scale and the general quality of the field staff used in them. Unless there are good grounds for believing that the quality of field work during the post-enumeration survey will be much better than that during the original inquiry, the post-enumeration survey cannot provide any useful standards for judgement, although it may provide some information about response variability.

There are other reasons why post-enumeration surveys have not proved popular other than after a census. They cost money—the cost per interview is usually much higher than that for the main enumeration itself, and it is tempting to believe that the money could be better utilized to increase the size of the survey sample. It may be difficult to check some of the survey data at a later date. Expenditures, consumption, and crop yields observed directly, or obtained with use of a very short recall period, cannot be checked directly against data relating to a different period, or obtained by using the identical reference period but with a longer recall period owing to the passage of time. We have argued that the success of a multi-visit survey depends on the development of a strong rapport between enumerator and respondent: a post-enumeration survey with a strange interviewer is likely to be of little use in such a case. A second interview on the same subject matter within a short space of time may be asking too much of a respondent, even in a one-off survey; is it fair to expect him to put up with, and to give equal concentration to, questions posed for the second time?

Another, although usually unspoken, reason for the neglect of the post-enumeration survey is that if it suggests not only random errors but also biases, the surveyor cannot see what he is to do. Does he abandon the processing of the survey or 'correct' the data for 'average' biases using the post-enumeration data? The ethical surveyor, once the post-enumeration data exist, will not contemplate suppressing the evidence of possible errors, but to publish conflicting results may seem uncomfortably like washing one's dirty linen in public to little useful purpose. In many cases surveyors know that resources are simply not available for carrying out the survey with the optimum methodology; the choice is between second-best and nothing, and they prefer to use what they have in making the main survey as good as possible.

A post-enumeration survey, if one is mounted, involves more than reinterviewing respondents. Checks of enumeration area demarcation and the coverage of respondents should be included. These aspects should have been checked before the census or survey was conducted, but it is still necessary to ensure that the enumerator has abided by the plan in keeping to the correct boundaries. This is particularly necessary in a census, where confusion in the boundaries may lead to under- or double-counting. If the design involves two-stage sampling, the listing of units and the sample selection should be made prior to the enumeration—the list of respondents is provided to the enumerator rather than allowing the selections to be made by him. The post-enumeration survey should ensure that the correct respondents were enumerated, and, if the final sample unit is the household, that the household definition was correctly applied.

The visit to the subsample of the respondents selected for the post-enumeration survey should be made as soon as possible after the actual enumeration to collect some or all of the responses to the survey questionnaire using an enumerator other than the one who conducted the original inquiry. One possibility, therefore, is to use the supervisors, but they should work in areas in the district other than those they were responsible for supervising. The subset of questions to be used in the post-enumeration survey should include those where careful probing by a skilled interviewer may uncover details that were not disclosed during the original interview.

The timing of the post-enumeration survey should be fixed at the survey planning stage, so that the logistics of moving supervisors into their post-enumeration areas can be worked out and the financing allowed for. The supervisor will not know the respondents in his assigned post-enumeration areas, so he must be able to locate them using the name, address, or plot number provided. The movement of the supervisors to different areas and the time it may take to locate the respondents selected for a revisit results in the higher cost per interview already mentioned. The supervisor should not be given the original questionnaires, and it should be made clear to him that the exercise is not a quick check of recorded data, but an independent collection of data.

The number of respondents to be revisited is a matter at the discretion of the surveyor. If the post-enumeration survey is very small, only impressionistic assessments will be possible. Considerable care should be taken to explain carefully to the respondents the reason why they are being troubled again, and why they should give the questions as much of their time and concentration as they did the first time. If the initial interview was of reasonable length and well conducted, the post-enumeration interview may be carried through without difficulty. If the initial interview was excessive in length or interrogatory in nature, we can scarcely recommend yet another equally trying interview to follow it; and it is not likely that the respondent will co-operate further. At the most, the post-enumeration in such a situation should consist of a small subset of the questions. The reinterview should be concerned with data that would normally be unchanged over the time period that has elapsed since the original enumeration. The recording of 'new' data, for example, expenditure during the previous day (a different day from that used originally), may provide very useful corroboratory material at an aggregate level, but can scarcely be used to test the accuracy of recording of the original data for that particular household—unless the differences are so extreme as to indicate gross enumerator negligence, even to the extent of faking the data. But such a procedure is not what we have in mind as constituting a post-enumeration survey. The use of interpenetrating samples and the like may be very useful for the detection of enumerator biases; 'new' data from a subsample of respondents comes more into this category than within the strict definition of a post-enumeration check on data quality.

Having gone to the trouble of collecting what purports to be identical data from a subsample of respondents, the next step is to carry out a matching of the data for each post-enumeration respondent. Simply to work out aggregate estimates from the post-enumeration survey and compare them with those from the original does not make full use of the potentialities of matched data. If this is all that is done, differences may be explained by the small sample size in the post-enumeration survey—a fault of the post-enumeration, not of the main survey. The difficulty is to decide what action to take as a result of an analysis respondent by respondent. Perhaps some enumerators are exposed as having been incompetent; possibly a systematic bias is detected such as an under-reporting of cattle; or the errors appear to be random but alarmingly high. Particular action must, of course, depend on circumstances, but some guidelines can be suggested. The examination of differences when reviewed across the entire subsample should cover both gross and net differences, or, to put it another way, the aggregates obtained both by ignoring and taking account of the direction of the difference. The average gross error is an indication of the size of the individual response error; the net error will give some indication of whether there is a systematic bias. The work of individual enumerators that is revealed as being of an unacceptable standard will have to be rejected, and the question of complete re-enumeration of the respondents concerned must then be considered. The treatment of systematic bias will be considered further in the next chapter—an adjustment at the aggregate level may offer at least a partial solution. It must also be remembered that although random errors may not seriously affect estimates of totals, disaggregated data resulting from cross-classification of variables may be subject to large errors.

The finest and most intensive supervision, together with the incorporation of a post-enumeration survey, will not ensure good quality data if the basic methodology of the survey is flawed. The enumerator can only be judged and held to account if he has been given the correct tools to do the job required of him. If he is asking the impossible or unanswerable question, as discussed in Chapter 6, the lack of consistency within the recorded data, and differences between the survey and post-enumeration data, may reflect nothing more than the respondent's inability to give the correct answer, or even to give consistent approximations to the correct answer. Conversely, consistency within the survey data and agreement between them and those of the post-enumeration survey may be falsely reassuring. If the recall period is too long the respondent may be forced into giving a rounded estimate that he will stick to under further or repeated questioning. Consistency may have been achieved, but not accuracy.

8.7 THE RIGHTS OF THE INVISIBLE RESPONDENT

The surveyor, when designing the survey, carries a major responsibility to the members of the public who will be the respondents; whether they are 'used' or 'abused' is largely within his hands. Unfortunately, many surveyors do not consider the eventual respondents as *persons*, but as final sample units. The surveyor may not come into much contact with the respondents; the respondent appears as an invisible anonymous being, waiting to be questioned at whatever length and on whatever topic the surveyor prescribes.

The respondent has rights—rights that the surveyor should not infringe, however inconvenient it is to avoid doing so. These rights include: (a) freedom from being asked personal questions about someone other than a very close dependant; (b) freedom from being subjected to enumerator behaviour that is uncouth, patronizing, overbearing, or threatening; (c) not having his courtesy, tolerance, and patience strained by excessive questioning and too frequent visits; (d) privacy, including the privacy of keeping certain personal information to himself if, he wishes to do so.

Exceptionally, such as in the case of a population census, the interest of the nation may supersede the rights of the individual as described above. But it is a common occurrence to find these rights being infringed by surveys that cannot claim this overriding priority.

It may be reasonable to ask the head of a household to answer questions concerning the way of life, occupation, and income of his spouse and children. It is not reasonable, in most instances, to ask the spouse or son to reveal these details about the head of the household if he is absent. Personal information of a sensitive nature should be given by the individual concerned or by the parent in the case of young children. The need to find and interview the correct respondent must be impressed on the enumerators. If this requires a recall at another time it must be so. The taking of the short cut of interviewing someone else about the respondent should not be condoned. The suggested core questionnaire for the World Fertility Survey[4] includes a section that deals with the background and economic activity of the respondent's husband. These questions include the following:

Can he read?

What kind of work does he do?

Does he get paid mostly in cash or mostly in kind?

Such questions infringe the principles outlined above. It may be found that in practice respondents will answer (or attempt to answer) these questions: but the answers will often be unreliable, and so what we consider rights will have been disregarded for little return.

We have already touched on the question of enumerator behaviour, particularly in Chapter 7. Too many enumerators feel themselves vested with authority that they do not possess, and adopt, if given the chance, a manner that the respondent should not have to tolerate. Unfortunately, it is likely to be the poorer, less well-educated or well-informed sections of society who will believe they have no alternative but to put up with such behaviour; so, inevitably, it is to these people that the enumerator will show himself in his worst light. Supervision, especially by the surveyor and his senior assistants, must pay particular attention to this point so that such tendencies are stamped out by reprimand and disciplinary action.

It is the surveyor who must be blamed if the length and number of interviews are excessive. Certain surveys demand frequent visits or lengthy interviews. In such cases the respondent's interest in the survey must be aroused so that he accepts, even if he does not welcome, the disruption of his normal routine. One of the ways of achieving this is to let the respondent see the results. If a household income and expenditure survey requires frequent visits over one month, the household members may be quite interested to know the details of how they spend their money, but if they are not shown the results they get nothing from the devotion of their time. It may be difficult to organize the supply of household summaries to the respondents, but it should not be an impossible task. Needless to say, great care should be taken to ensure that such household summaries do not fall into the wrong hands.

Similarly, if the holding of the household is being measured, together with the fields and plots, the farmer who has to guide the enumerator around and possibly help to carry the necessary equipment may like to know the precise areas he has under cultivation. This should not be too much to ask. The summary information should, of course, be supplied at the end of the survey. If provided earlier, the respondent's behaviour may be changed during the survey by the information supplied. The surveyor should also bear in mind that to keep a respondent in a continuous survey for a long period may affect his attitudes and behaviour, thereby making him untypical. Not only has the respondent been excessively 'put upon', but the results may be biased.

Payments to respondents set dangerous precedents, and may affect the survey process. We recommend that they should not normally be made except, as already suggested, when a respondent agrees to undergo a series of interviews in order to help train enumerators. In these circumstances he becomes similar to a part-time assistant.

Many survey interviews are needlessly long. The inclusion of non-essential matters in the questionnaire and the wish to collect data at a level of refinement that is not required are the main causes. Some interviews are inevitably long because the survey is designed as a multi-subject integrated enquiry—the Living Standards Measurement Survey, mentioned in Chapter 4, is an example. Evidence on the declining quality of response due to fatigue of the respondent is scanty. But we remain primarily concerned that advantage is sometimes taken of the tolerance of the population who may not feel themselves to be in a position to complain:

we find it difficult to believe that some of the interviews, as currently practised, would be accepted by households which are willing to voice their objections.

The extent to which respondents are bound by statute to answer questions depends on the statistics act or its equivalent in the respective country. Most surveys are not conducted with the force of statute and the respondent has the right to refuse to co-operate. Fortunately, given proper explanation, the incidence of non-response will be low, but the respondent is due this explanation, and he is right to expect that essentially private details will not be investigated for trivial reasons. Above all, he has the right to expect that the information will be treated as confidential. Assurances on this point are freely given, but not so rigidly observed, particularly after the survey has been completed. The provision of personal particulars to anyone for whatever purpose should be rigidly prohibited. The destruction of questionnaires and the removal of identifiable household particulars on the stored data should be a much more common procedure than it is.

We have spoken about the need for the interview to be held in private in Section 8.3. Mature sons and daughters in the household may not wish to discuss details of their own lives in the presence of their parents or other senior relatives in the household. In many societies it may be almost impossible to arrange to interview the teenage girl without the presence of her mother or mother-in-law. If a private interview cannot be arranged, the desirability of the enumeration should be seriously reviewed by the surveyor. This will not be a popular view with statisticians, and we have not always adhered to it ourselves: we are, however, becoming alarmed at the growing tendency of surveyors to assume they have *rights* to any information. The need to know must be pressing indeed to justify the embarrassment that is sometimes inflicted.

Notes

1. Sahib, M. A., *et al.*, *The Fiji Fertility Survey: A Critical Commentary—Appendices*, Occasional Papers No. 16, World Fertility Survey, London, 1975, Appendix 2, p. 8.
2. Sahib, M. A., *et al.*, *The Fiji Fertility Survey: A Critical Commentary*, Occasional Papers No. 15, World Fertility Survey, London, 1975, p. 53.
3. Ibid., p. 43.
4. World Fertility Survey, *Core Questionnaire*, Basic Documentation No. 1.

9

Data Preparation and Processing

I examine the data, as an expert, and pronounce a specialist's
opinion.

The Sign of the Four

9.1 GENERAL

The title of this book refers to data collection, and our main focus is
naturally on problems connected with this aspect, rather than on the
processing and analysis of data. But, if a survey is to be successful, the
required output must be specified in advance and the uses to which the
data are to be put must be explored. For this, if for no other reason, it is
appropriate to devote this and the next chapter to the steps required to
complete the survey, subsequent to the data collection.

Data collected in many surveys in developing countries have never been
analysed, or, at least, the analysis has never been completed. Results from
many more are so delayed in production that much of their value has been
lost by the time they are published. The record of timely analysis is better for
smaller, usually non-governmental, surveys, as might be expected. The fault
with many of the reports from these smaller surveys is that the analysis is
deficient in quality and depth because of poor data processing arrangements.

Not all the blame for the failure to produce survey results quickly can be
laid on indifferent data processing. Data processing is an integral part of
the survey operations and is affected by its overall quality. If the
questionnaire and survey methodology used were faulty, very little
effective data processing may be possible. Nevertheless, there has been a
tendency for survey designs to be such that, although adequate in
methodology and content, and resulting in data of reasonable quality,
they are not brought to a successful conclusion. What was wrong? While
there is no single answer to this question, the most commonly encountered
contributory cause is failure to plan the processing. Many months are
spent organizing the collection of the data, yet little or no thought is given
to their processing. With the arrival of the completed questionnaires it is
realized, too late, that the necessary processing facilities in terms of men
and machines are unavailable, or only partly available. Loss of momentum causes delay, which leads to loss of enthusiasm and possible reallocation of staff to apparently more pressing issues, with consequent further
loss of momentum, and so on in a depressing spiral. Moreover, we have

noticed that some surveyors have a peculiarly blinkered outlook that leads them to regard the collection of data as an end in itself. Rather than concentrating on the preparation of a report when the field work has been completed, they turn their minds to designing the next, leaving the processing and writing-up in the unsupervised hands of junior staff, inadequate both in number and experience. Surveyors engaged in evaluating projects stated at a recent workshop held in Africa that one cause of processing delays was a 'tendency to abandon analysis and interpretation in favour of continued field work',[1]

Sometimes, problems encountered with part of the data sour the surveyor's attitude to the remainder. The survey may have failed to collect good quality data relating to some variables, but may have succeeded for others. Partial success is better than total failure; indeed, it may be all that could reasonably have been expected. Moreover, disappointment that the data are not suitable for sophisticated analysis, such as hypothesis testing or model building using advanced statistical techniques, should not deter the surveyor from presenting the data in a simpler form. Such a presentation may still be of great value to many data users, and may serve to indicate that concentration on the original objectives of proving certain a priori hypotheses was misguided as relationships betwen key variables were more complex than expected and need further investigation.

9.2 PROCESSING PLANNING, PRIOR TO DATA COLLECTION

We have already stressed the need to consider data processing requirements at the time the survey is planned, arguing that failure to do this is the main cause for excessive delays in producing survey reports. There are several questions related to data processing that must be answered at the survey design stage including the following:

(a) What is the output to be?
(b) How is this output to be produced?
(c) How should the data be recorded in order to facilitate the required processing?
(d) What resources are required to achieve this processing speedily and accurately?

When the survey objectives are being discussed, the required output should be visualized in a general form. If the objective is description, the output will be mainly in the form of frequency distributions and cross-tabulations: if the objective is to examine the interrelations of many variables then a more analytic output will be necessary, imposing requirements on the data in form, structure, and quality. Most surveys will involve both description and analysis; but as the latter is more specific to the individual inquiry and requires a detailed matching of data to a range of sophisticated techniques, we shall concentrate on description, and that but briefly. Surveyors should consult more detailed treatment of these issues in books such as that by O'Muircheartaigh and Payne.[2]

After agreement on the broad outlines of the form of the output has been reached during the user–surveyor dialogue, detailed decisions should be taken concurrently with the design of the sample, the choice of the methodology to collect the data, and the design of the questionnaire. That the required output should now be fully specified is an essential requirement for efficient design; at this stage, vagueness will not do. No methodology or questionnaire should be given final approval without consideration of the tabulation programme to be related to it. If a general statement about the extent of literacy is required, a few simple questions regarding schooling and a simple probe of the respondent's claim to be literate may suffice. If it is desired to measure the extent to which those who had received some primary education are now illiterate, a careful test of ability to read and write is necessary, for few people will admit to having once been literate but to being so no longer. Precise questions on the extent of primary education must also be included. The resulting tabulation requirements will be very different in the two cases.

The specification of the output should include the following:

(a) A draft set of tabulations. Precise definition of the size groupings of a variable for which frequency distributions are required may on occasion be left until the time of the analysis, but the variables to be tabulated should be clearly indicated (but see (d) below). This exercise will reveal unnecessary inclusions in the draft questionnaire, and the unfortunate omissions.

(b) The hypotheses to be tested and the relationships to be explored. The formal statistical analyses involved impose requirements and constraints in the data collection process that may be overlooked unless the analyses are specified in advance. It has been claimed that the conditions under which tests of significance can be used are not met in survey work.[3] A more optimistic assessment, accepting the lack of control, stresses the need for 'judicious collection and use' of data.[4]

(c) The draft contents of the intended report. This is very rarely done, but often, when a report is being written, it is realized too late that in order to show that the report's conclusions are valid a related topic for which no information exists has to be discussed.

(d) Information for comparative purposes. It may be possible to produce the data in a form that enables comparisons to be made with earlier surveys, or with data from alternative sources. If so, the desired output should be carefully specified so that the data collected is grouped as required. Here again the question of setting the findings of the survey within the general framework of its subject needs consideration. Some idea of the extent to which this will be required will be obtained when the need for the survey is being examined (see Chapter 2).

Once the output has been specified, the choice of methods for producing this output must be made, although in practice the choice of output and the method of producing it cannot be separated in this precise fashion. Experience with many surveys shows that computers do not speed up the production of essential tabulations—they do, however, facilitate the production of a comprehensive set of cross-classifications of key variables that it would not be feasible to undertake using calculations alone. It is when the volume of data is large that direct analysis by the surveyor becomes impracticable, although there is still a great advantage in analysing at least a sample of one's own data. This personal 'handling' of the data gives insight into their idiosyncrasies, limitations, and potential. It will facilitate the report writing and assist the surveyor to avoid drawing erroneous conclusions from computer print-outs that, by summarizing and aggregating the data, may conceal vital evidence or confound important relationships. Failure to follow through in this way is one reason why a surveyor's abdication of responsibility for the data analysis, mentioned above, can have such a fatal effect on the final output.

In the normal case, then, a computer will be used for at least part of the processing. When mainframe computers are used, the surveyor is usually somewhat distanced from the data processing as these computers are rarely under his direct control. However, the advent of microcomputers allows the surveyor to retain a close contact with, and responsibility for, the data processing.

If one or another type of electronic processing is to be used, the data must be in a form that is acceptable as an input into the computer. A questionnaire with answers recorded in a coded format by the enumerator is of course the ideal, but this most economic solution is not always possible. If not, it is important that the recorded data should be in a format that can easily be transformed into a suitable code.

This is where the open-ended question with written answers that are not easily coded becomes a problem. By some means these answers must now be classified into a summary code. It might well have been better to have researched the problem and to have adopted an appropriate summary of possible answers at the questionnaire design stage, so that the data recording could be structured accordingly. This research is a suitable topic for a pilot study. The popular last resort option of 'other, not elsewhere specified' can be used for the cases that do not fit into the summary answers finally provided, although it is a confession of failure if many cases have to be put into such a group. The small survey and the case study allow greater flexibility: when the total number of questionnaires is limited each can be examined and summarized by the surveyor or a close associate.

Other references to data processing convenience in the context of questionnaire design have been given in Chapter 6. One further point deserves mention: if the data from one survey are to be linked with those from another, reference codes of the linking unit must be compatible in both sets of data. Apart from the obvious need to keep, say, the household code number fixed for all questionnaires relating to that household, the identification of children on a questionnaire relating to nutrition, for example, should be such that they can be linked with their natural mother on the household listing questionnaire. In a multi-subject household survey the need to cross-reference data on different computer files frequently arises in the second phase of analysis. It is not uncommon to find that such a process is either impossible or impractical because of deficiencies in recording the original data, deficiencies that could easily have been remedied if analytical needs had been considered in detail at the survey design stage.

The resources required to achieve quickly the necessary data processing will almost certainly be underestimated at the planning stage, with regard both to quality and quantity. The surveyor must usually accept responsibility for the preparation of the data right up to, and including, the preparation of discs or tapes. He will almost always have to organize the staff to handle the editing, coding, and validation of the data. If a computer is not being used, staff will be required to perform the more routine arithmetical and statistical computations. The creation of an efficient processing team in good time before the arrival of data from the field is as essential as the creation of a good enumerator team. For a national sample, generating a continuous stream of data, the number of analysis staff required will be approximately equal to the number of field supervisors unless the questionnaire is extremely simple and pre-coded.

If a computer is used, it is necessary to ensure that it will be accessible at the proper times, that the appropriate software packages are available, and that the staff have been trained in the necessary skills. Inadequate attention to these preparations will result in frustration later when the edited data are ready for processing. Other work, then accorded higher priority, may be occupying the input preparation facilities, or the required statistical package may not have been compiled on to the computer operating system. If the required software is not available it must be tailor-made and debugged in good time. It will be too late to commission fresh software when the enumeration has been completed.

Rapid proliferation of microcomputers has commenced, even in the poorest countries. A choice of the most suitable hardware for survey data processing is often precluded either by government policy to purchase only from certain manufacturers or because servicing facilities exist only for certain makes. Of much greater importance, however, is that the software not only be compatible with the hardware available, but optimal for the tasks required. Only a very few years ago the problem was that good software was unavailable: the problem now is that such an enormous variety of packages exist that it is difficult for a surveyor, remote from the countries of origin, to know which are best suited to his purpose. The range runs the entire gamut from data editing to advanced multivariate statistical analysis. They are produced by universities, international agencies, and commercial software firms. Prices range from a few hundred pounds to many thousands, depending mainly on the range of features provided and their ease of utilization. Once again, if the tabulation and analysis specifications have been set down it should be possible for the surveyor to obtain advice on the software suitable for his needs. It should be borne in mind that any one package is not likely to meet all his requirements. For example, one package may be needed for primary data manipulation and another for later statistical analyses. With the range of software now available we advise against attempts to write tailor-made programs for a particular survey. It is not easy to rival the efficiency of the standard packages, nor to debug such home-made efforts.

Microcomputers are simple to operate, relative to former computer generations—they are user–friendly, in the jargon of the trade. Nevertheless, the surveyor requires staff who are familiar with both their general operation and the software packages to be used. Some programming expertise may also be necessary to handle *ad hoc* needs, previous comments notwithstanding. If the data files are eventually to be transferred to a mainframe computer, special skills are required. A common problem, which often presents itself unexpectedly to surveyors, is that various software packages have different requirements for data input formats and some of these do not provide the 'utility routines' to transform data from another format to the required one. This can be a major cause of delay unless staff with the skill to write such routines are available.

One further possibility also exists as a result of the availability of cheap microcomputers, namely the use of such computers in the field during the data collection phase to provide direct data validation and a data record that obviates the need for questionnaire handling in the main survey office. Use of microcomputers in this way has been pioneered in a few surveys using a very small number of mobile teams, but experience is limited, particularly where the computers must tolerate harsh environmental and travel conditions. Nevertheless, it can be said that such a practice is not as infeasible as many imagine, and mobile teams armed with such an aid are likely to grow in popularity. Clearly, in this, as well as office applications, there must be adequate provision for back-up hardware, spare parts, and maintenance facilities.

9.3 DATA PREPARATION

The preliminary phase of data processing includes the following:
(a) reception of the data;

(b) editing and coding;

(c) initial examination of a sample of the data;

(d) data entry.

The questionnaires for a survey operating on a wide geographical base arrive at the survey office from the field by various means and in irregular amounts. Some may be hand carried by a survey officer returning from a field visit, the rest will come by post, or through private parcel transportation and delivery systems. If the survey is based on a single interview, conducted within a short time period, the questionnaires will arrive in bulk; whereas questionnaires from a cyclical survey will arrive in an almost continuous stream. Whatever the mode and frequency of arrival, it is important that the questionnaires are booked in and sent to the start of the processing system in a foolproof and orderly manner. Questionnaires may get lost in transit, but they can also be lost within the analysis office. If the record of arrivals and transfers is casually or vaguely maintained, it will be impossible later to check the whereabouts of missing data or to attribute responsibility for this state of affairs. If some questionnaires fail to arrive from the field, this must be noted quickly in order that a query is sent to the appropriate supervisor without delay. Belated queries may fail to produce either the original data or a repeat set, whereas a prompt follow-up will reduce substantially the amount of missing data that give rise to subsequent problems.

The office record of questionnaire flow should identify each enumeration area, or even in some cases each respondent, and show the dates on which the relevant questionnaires arrive at, and are dispatched from, each section of the office, commencing with the arrival of the questionnaires from the field. Questionnaires are best moved through the stages of data preparation in batches, preferably by enumeration area, and should be filed or otherwise secured, to prevent detachments. A record of data flow is equally important at a later stage when tabulations are produced. A set of computer tabulations may be divided into sections for further work by various analysts. It is wise to maintain a master file, which contains a copy of all the computer tabulations, and to keep it as a reference 'library'. Many useful tables have been lost to later use because no master file, in the form of either tape or print-out, was maintained.

The role of the editor has been discussed in Chapter 7. We repeat that, even if a comprehensive computer validation is to be conducted, there is need for a data edit immediately on receipt of the questionnaires from the field. The last opportunity to *replace* errors with correct information, as opposed to *cleaning and imputation*, eliminating and adjusting dubious records, may be at the time of the edit with follow-up field checks where necessary. The objectives of the edit in sequential order are:

(a) to detect data that are incorrect owing to apparent misunderstanding by enumerator or respondent and return the relevant questionnaire to the field for correction;

(b) to detect sets of questionnaires containing data superficially acceptable, but which on closer examination raise suspicion of 'faking' by an
enumerator;

(c) to correct obvious misclassifications or recordings that are mere 'slips of
the pen'.

At this stage, it is worth conducting an examination of orders of
magnitude for selected variables and making a first estimation of key
ratios. The editor should extract the selected data on to a simple record
sheet in order that these estimates can be made. Whether the extracts are
taken from each questionnaire, or from only a sample of them, is, of
course, dependent on the size of the survey. Apart from the intrinsic value
of this exercise, it also provides the editor with a specific task that assists
concentration when examining the questionnaires. The extraction of
information should not absorb too much of the editor's attention and
replace the normal editing checks but remain, rather, additional to them.

Such cumulations of the data for the selected variables provide the
surveyor with quick estimates of totals that will be produced from the
survey, and estimates of ratios. The variables selected should be those for
which knowledge concerning their likely orders of magnitude already
exists. For example, from previous surveys it may be known that the
number of cattle in the country is of the order of 5 million. A running total
of cattle counted on selected holdings in an agricultural survey enables
preliminary calculations to be made of the likely survey estimate. If a final
total of the order of 10 million is indicated, the surveyor is provided with
an early warning that something is wrong with either his sample or the
provisional count of cattle. There may still be time to find the source of
error and take appropriate measures to counter it. Similarly, in an
anthropometric study of young children, one key ratio is the child's weight
for height. An early examination of a sample of weights for heights may
confirm that the survey has been conducted with the required accuracy.

The detection of cleverly faked data is one of the hardest, but also most
important, tasks of the editor. The main responsibility for preventing
enumerator dishonesty rests of course with the field supervisory force; but a
skilled edit may detect evidence that was missed by the supervisor. A quick
scan of all the questionnaires completed by one enumerator may reveal a
suspicious homogeneity in the data: in an expenditure survey each
respondent buys milk every day; all households contain a 'set-piece' family
with only minor variations; there is a suspicious absence of occasional zeroes
or not specified. The expected departure from the norm does not occur and
the unusual is non-existent. This repetition of the commonplace is one of the
distinguishing marks of faked data—which is not to say that all such data
are faked, but rather that most faked data will exhibit this feature. The edit
of a particular enumerator's work should concentrate on the detection of a
pattern which, if present, should be checked by a field inquiry.

Errors resulting from misunderstanding or neglect of duty by one or several enumerators can often be detected by comparing the proportion of some subset of the population obtained from each enumerator. Deming suggests plotting for each enumerator the count of a specified subset of the population against the count of the remainder of the population using binomial probability paper.[5] If one or more points are well removed from the others the work of the enumerators concerned should be checked. If membership of a particular group involves an enumerator in extra work (for example, the completion of an additional form), then the proportion in this group, or those falling just outside, should be watched.

If computer processing is not involved, the next stage consists of the extraction of the data from the original records on to summary sheets incorporating frequency distributions, two-way tables, and totals. Cross-checks and control totals are required. Operations of this kind are paralleled by the computer, but on a different scale: the extra stages introduced if computer processing is involved, are considered next.

Guidelines, which should be observed at this stage, are:

(a) all codes should be numeric;

(b) the identification code should identify uniquely not only the respondent but also his location in terms of geographical area, and, where appropriate, the sample stratum and/or cluster;

(c) a distinction should be made between a response with the value zero and a non-response;

(d) totals that can be calculated by the computer from the individual component values should be entered only if a check with the computed value.is to be used in validation.

By this time all the necessary systems design, data dictionaries, and computer programs for the production of tabulations should be ready, and decisions made regarding the use of particular software packages. If the analysis procedures are completely specified, the systems analyst can organize the data file so as to facilitate subsequent processing. In a continuous survey, requiring the creation of a sequential record to which data are added after each survey cycle, the surveyor must brief the systems analyst on his requirement in terms of final tabulations.

Before proceeding to the main data processing, it is necessary to validate the data as now recorded on disc or tape and to 'clean' the data file. This operation is essential. Despite all the checks in field and office, errors will be present, errors that may seriously affect the validity of the resulting estimates. A validation program is required that will detect erroneous data and print the offending record for manual checking and correction. It should include the following checks:

(a) the number of cases for each variable on the file must agree with specified totals;

(b) all codes must be within the specified range;

(c) specified consistency checks must be satisfied;

(d) numeric values that lie outside a specified range must be recorded.

Bailar and Lanphier refer to one major survey organization that 'performed no machine edit unless "something looks funny in the tabulation"'.[6] This practice cannot be recommended. To ignore validation is foolhardy; on the other hand, it is possible to spend so long seeking a 'perfect' file that major delays are introduced. We recommend that no more than three error print-outs should be produced and corrected by the editing and coding staff. Each stage will produce further errors and if the process is not stopped our experience shows it can continue for six or more validation runs. After two or three sequences of error correction a program should be used to substitute imputed values or eliminate remaining errors by the application of simple rules dealing with permissible code sequences, and the range within which a value must lie. The purpose is to produce valid, but not necessarily perfect, results in as short a time as possible; not to indulge in a long-drawn-out attempt to correct a few remaining errors which, if 'eliminated' from the file, will not materially affect resulting estimates.

At some stage during this process the problem of non-response must be dealt with.

The results of the survey will invariably include estimates of the effect of nonresponse, whether these are implicit or explicit. The typical situation is one in which the extent of the nonresponse is reported and the results are calculated on the basis of the responses. This approach implicitly assumes that the nonrespondents are sufficiently similar to the respondents to justify ignoring them in the calculations. It is equivalent to treating the achieved respondents as a representative sample from the population.[7]

This quotation from O'Muircheartaigh prefaces a discussion of other procedures that may, and often should, be adopted. The surveyor should be familiar with the procedures of reweighting, interpolation, and the use of a model to predict missing values. A simple option is known as the hot deck procedure, where a case with similar characteristics to the missing one is selected at random and duplicated as a substitute. This may be useful to complete partial response. A cautionary word on substituting for non-response is appropriate. Kish states that, 'no method of substitution is generally free of disadvantages. These often outweight the advantages. . . . But we may choose the method with least disadvantages for a specific situation.'[8] The surveyor must be clear about the assumptions that explicitly or implicitly lie behind the procedure adopted: the relevance of these assumptions determines the quality of the procedure. Thus the first approach mentioned above, which assumes that non-respondents are similar to respondents, although simple, is usually inadequate.

Once a 'clean' file (with or without induced elimination of errors) has been achieved, the data should be protected by using a *copy* of the file for processing, keeping the 'master' in separate storage: accidents can and do happen to discs or tapes when they are being read by the computer. When file security is assured, the confidentiality of the original data should be guaranteed by removing from all records the means of identifying individual respondents.

9.4 ESTIMATES AND THEIR ERRORS

The main processing stage is the production of estimates from the survey data, whether these are totals, means, or frequency counts. In order to produce these figures, the first issue to be resolved, if the survey is based on a sample, is the weights to be used to rate up the data. If the number of respondents covered was small, from a limited geographical area, and selected with unknown probabilities, the survey is a case study or a series of case studies, and the data should be presented as representing the selected respondents only; they should not be weighted up to estimates purporting to represent a wider population. If the survey is based on a valid probability sample, estimates of population totals can be given.

The calculation of approved estimates results from the following processes:

(a) the rating-up of the sample values to population estimates;

(b) the evaluation of biases and other non-sampling errors;

(c) the use of supplementary information to improve the estimates;

(d) the calculation of sampling errors;

(e) the setting of the levels to which the estimates can be disaggregated in tabulations;

(f) the use of standard statistical packages for the production of the tabulations.

The use of the weights calculated from the selection probabilities is an easy enough matter in principle. The formulae for producing totals, means, etc., from a sample of specified design are well described in all standard texts on sampling theory. Unfortunately, it sometimes occurs that these estimates, when calculated for a few basic totals for which other estimates exist, appear grossly incorrect. If the estimated total is of a variable such as income, it may well be that income has been misstated by the respondents—that is the estimates are biased owing to individual errors. But if the estimate of the size of the population or the number of households does not accord with other knowledge it is likely that the error is inherent in the sample selected, a fault in the execution if not in the sample design. This problem particularly occurs when the sample design is complex. In such a case, the calculated probabilities of selection of the respondents may be wrong.

Reasons for this divergence in a multi-stage clustered sample may include:

(a) theoretical sample design faults;
(b) faulty mapping of selected enumeration areas;
(c) faulty listing of the final sample units in selected enumeration areas;
(d) biased techniques for dealing with non-response, vacated dwellings, etc.

Kish reports that 'practitioners believe that in most social surveys noncoverage is a much more common problem than overcoverage. . . .'. He reports his belief that '10 per cent noncoverage for national samples is often exceeded'.[9] In their assessment of USA practices, Bailar and Lanphier report, more generally, that '15 of.26 federal surveys did not meet their objectives, four because of poor design, four more because of failure to implement plans for probability sampling, and the remaining seven because of a combination of serious technical flaws'.[10]

We have discussed some common faults in sample frames in Chapter 4, and also errors committed in mapping, listing, and implementing the design. Non-response, and biases introduced in dealing with it, can take many forms, including simple substitution of clusters that are inaccessible or otherwise difficult to enumerate; the assumption that vacant dwellings at the time of listing are permanently vacant; and the toleration of the loss of respondents during the survey without adjusting the weighting procedure. Biases introduced during analysis by the treatment of non-response may still be remedied by correcting the procedures, even at this late stage. But, in general, methodological faults in the survey cannot now be corrected.

Supplementary information can sometimes be used to adjust estimates obtained from the sample. Independent estimates may exist for the population of each small administrative area; for example, census figures adjusted for any time lapse that may have occurred. Enumeration of a sample of these areas in the survey may have been subject to errors caused by faulty mapping and listing. The simple estimate of the total population obtained by multiplying the sample total by the reciprocal of the sampling fraction may have to be replaced by one obtained by rating up the sample total concerned to some appropriate accepted population figure. Hansen, Hurwitz, and Madow[11] give an example of this procedure used in a survey of Haiti. The use of ratio adjustments (which are to be distinguished from ratio estimates) will result in figures that are themselves biased, but they may be more accurate than the use of apparently unbiased techniques, which proceed as if no survey errors had occurred.

Adjustments for bias can be made at the level of cluster, stratum, or region. The choice of procedure depends on the assumptions made regarding the nature and location of the bias. Post-survey investigations may reveal that certain clusters were wrongly demarcated in the field leading to under-counts. Adjustment at the cluster level will be appropriate in such a case. On the other hand, there may have been a general tendency to under-list the households in each selected cluster: the bias is distributed throughout the sample and the adjustment should be likewise. If the biases are inconsistent from one variable to another or are due to incorrect responses, ratio adjustments will not generally improve matters; indeed, they may compound the problem. Cash income from casual employment may be grossly under-reported, whereas crop yields may be exaggerated. Little can be done to repair these enumeration incurred biases.

If the problems associated with weighting the sample data are particularly serious, estimated totals for the universe under study may have to be omitted from the report. Something of value may still be salvaged from the data. Ratios of key variables are often not very sensitive to differential weights, so that selected statistics derived from the unweighted data may be used.

It should be emphasized that recourse to such procedures should be a measure of last resort. Care during the survey design stage, with proper allocation of resources, should prevent most of the problems. The sample procedures should be tested before the survey, not at the data processing stage. In a two-stage cluster sample, a trial weighting of the households listed as a basis for cluster frames will enable design or mapping faults to be detected and corrected in good time.

Although we have concentrated attention on the non-sampling errors that arise in surveys, the sampling error must be given due attention. Certainly the processing of survey results should include the calculation of sampling errors for a representative selection of variables at different levels of disaggregation. The use of computers has taken the tedium out of these calculations and reduced the need for short-cut, simplified methods.

The calculation of sampling errors, together with consideration of the size of non-sampling errors, will enable the surveyor to decide on the level of disaggregation that he can use in the tabulations to be released for use. A national sample of several thousand respondents may provide estimates of totals at the regional level within acceptable small confidence intervals. But separate estimates by income group for each region may be subject to sampling errors that indicate their unreliability. This poses a problem for the surveyor, since he will find himself under pressure to provide tabulations at a very disaggregated level for users who wish to examine one district in relation to another, or one income group against another. These users may not appreciate, and will often deliberately disregard, the size and implications of the error margin. Most users will not refer to any account of errors each time they consult the tables. Once printed, the figures will be quoted, used, and abused with no reference to their reliability. The surveyor can hardly be blamed for this, but it suggests that he should be cautious about the level of disaggregation published.

Surveyors must use packages if they are to obtain full benefit from the computer. There are dangers here, particularly when multivariate analysis packages are involved. These usually manipulate the data in complex ways, which nevertheless often results in an apparently simple final output; but the input may not be suitable without a transformation or other adjustment. A surveyor who uses these packages to extend his analysis into areas in which he is not already competent runs a serious risk of producing and reporting incorrect findings. Fortunately, the situation with regard to packages for tabulations is not so perilous. The surveyor may find he has a problem of choice. If so, he should consult a specialist and take account of:

(a) portability, that is the ease with which the program may be transferred between different machines;

(b) the clarity and fullness of error diagnostic messages;

(c) the levels of documentation, particularly the lucidity of the guide for non-specialists;

(d) interactive facilities;

(e) the need for a package which will handle 'multiple-response' answers in a convenient form;

(f) the advantage of packages which provide facilities for handling hierarchical data.

One desirable feature of a package is that it should produce good tables. A distinction can be drawn between what may be termed primary tabulations, which will be further edited and condensed by the surveyor into tables in the survey report; and final tabulations, which will be printed in their original format, or circulated directly to users. As sophisticated users know that a computer and supporting 'software' packages are available, the issue of a good survey report stimulates requests from individuals for further tabulations on a specially commissioned basis. Even the one-man private researcher will find himself receiving 'orders' for tabulations from his survey. The existence of a well-designed data file and suitable programs means that many of these requests can 'be met. If the user is to be given the computer tabulations with little or no accompanying text, caveats, and footnotes, it is particularly important that the tabulations should have the following characteristics:

(a) layout clear and pleasing to the eye;

(b) understandable by someone not intimately connected with the data on which they are based;

(c) valid in that the proper levels of disaggregation have not been exceeded;

(d) incorporates only those statistical analyses appropriate to the source data.

The above requirements need not be so rigidly observed when the tabulations are primarily for the surveyor. He, presumably, understands his own data and can follow his chosen path through even the most maze-like tabulation. Nevertheless, such a surveyor will still find that his analytical tasks are made much easier if he has clear tabulations in front of him. And it is surprising how often, when using data output, the surveyor conveniently pushes to the back of his mind the imperfections of the source data that were dominating his thinking a few weeks earlier. The surveyor should remember, too, that he may have to return to his tabulations at a later date, when his memory of the data characteristics has faded, only to find that the tabulations are not documented in sufficient detail to allow their further use.

Yet again our philosophy can be summed up in the words: keep it simple. Tabulations that involve more than three tabulating variables become very difficult to follow. Some software packages are better than others in producing tabulations that are concisely constructed.

The choice of the tabulating variables and class intervals to be used needs careful consideration, whether for frequency distributions or for cross-classifications. If the variables are discrete or non-numeric in nature, such as household size or region, there is a need to group classes if there are too many, or if the number of respondents in some of them is too small. For example, if there are twelve regions and the tabulation is by number of cattle, there may be so few cattle in three of the regions that the estimates are negligible or unreliable. Is it appropriate to group these three, consisting of, say, a coastal region, a northern hill region, and a southern desert region, into one class of 'All Others'? Or should each of them be linked with its immediate and better populated (in terms of cattle) neighbour, thus introducing three composite regional classifications? The choice will depend on circumstances, but a 'rag-bag' group is of little value to users, so the second alternative may be preferred. If grouping is necessary, the footnotes to the tabulation should include the total number of cases which have been amalgamated into a composite group.

If the variables are continuous, or practically so, the choice of class intervals is not always obvious. There are several objectives which may not be easy to reconcile, including the following:

(a) The boundaries of each class should be 'sensible'. A class for those households with incomes between 50 and 120 monetary units leads the user to wonder whether there is some significance in the 120. Why not 100, 125, or 150? Is 120 the accepted poverty line?

(b) Class intervals should be constant for as many of the classes as is practicable. It may be convenient to have wider intervals at the upper end of the distribution, but variations should be kept to a minimum. If a logarithmic transformation is to be used, intervals that are multiples of the standard class interval will be desirable.

(c) The number of respondents in each class should be sufficient to justify showing that class separately, and where varying intervals are used, about the same number of observations should appear in each class.

9.5 RETAINING CONTACT WITH THE DATA

Earlier in this chapter we emphasized that the surveyor who relies on a computer and computer packages for data processing may lose contact with the data. All he sees are well-ordered computer print-outs that in their very neatness and detail give an impression of solidity, which may be illusory. This loss of contact can lead the surveyor to accept, publish, and grant the authority of his reputation to analyses that are either misleading or false. In addition to editing a sample of the data himself, the surveyor should, following the creation of a computer file, obtain detailed print-outs for at least a sample of the data before summary tabulations are produced. If the surveyor summarizes and analyses the sample distribution himself, he will soon be well aware of the idiosyncrasies of his data; he can then handle the final tabulations with greater assurance that a serious error of interpretation will not be made.

It may be thought that these recommendations to do oneself what the computer can do in a fraction of the time are old-fashioned: the computer can be programmed to indicate that data are skewed or heteroscedastic; outliers, that is values outside the normal range, can be removed, and so on. Yet we would still insist that a personal examination of individual records is essential. The validation program searches for abnormality, but an examination of the individual records may reveal a peculiar pattern of excessive normality. Survey data are not laboratory readings on an accurate instrument. The respondent, enumerator, supervisor, editor, and coder have all manipulated the data. Some of this manipulation will have introduced distortion. Even if all is well and the data are 'clean' and of high quality, there is still an extra insight to be obtained from perusing individual records. Patterns and relationships may be sensed that will lead to worthwhile adaptations of the tabulation and analysis plan. Obvious lack of relationship may save much time running sterile correlation matrices. The existence of groups of data each having a distinctive distribution may throw great light on the main objective of the survey.

The personal examination of the data file can take many forms, including manual arrangements into frequency distributions and the search for outliers. These may often be usefully discussed with colleagues with local knowledge. They can throw light on the general background, and in small communities may be able to contribute in detail. Thus, in a survey investigating the relationship between farm training and subsequent performance, the local agricultural officer could account for many apparent outliers; for example, identifying a case where the husband had been trained but the wife was doing the farming. Another useful aid is graphical analysis.[12] Many unwarranted analyses would never have been undertaken if the surveyor had looked, even casually, at a graphical representation of the data.

Another useful general approach which often highlights distinctive features of a situation is to consider the following equation:

Observation = Value fitted by the model + residual

In its simplest form the value fitted by the model may be the average (or median) and the resulting residual for any observation would then be its deviation from this average. An examination of the series of residuals at the preliminary analysis stage will identify outliers which require further study. If the residuals are put in some order—by size, geographical or chronological sequence—valuable insights may be obtained. A series of models of increasing sophistication can be used to produce the fitted values, and a study of succeeding sets of residuals· can lead to more effective analysis. Further details can be found in Ehrenberg and Tukey.[13] The observations studied may be the original observations or a transformation of them—their logarithms, square roots, or reciprocals. As with graphical techniques, the basic idea is so straightforward that it can also provide an effective method of presentation, the subject of the next chapter.

Notes

1. IBRD, *Draft Report of Regional Workshop on Monitoring and Evaluation of Rural Development Projects in Eastern Africa*, Washington, DC, April 1979.
2. O'Muircheartaigh, C. A., and Payne, C., *The Analysis of Survey Data*, 2 vols, Wiley, London, 1977.
3. Selvin, H., 'A Critique of Tests of Significance in Survey Research', *Amer. Soc. Rev.* 22, 1957.
4. O'Muircheartaigh and Payne, op. cit.
5. Deming, W. G., *Sampling Design in Business Research*, Wiley, 1960.
6. Bailar, B. A., and Lanphier, C. M., *Development of Survey Methods to Assess Survey Practices*, American Statistical Association, 1978.
7. O'Muircheartaigh and Payne, op. cit.
8. Kish, L., *Survey Sampling*, p. 558.
9. Ibid., p. 529.
10. Bailar and Lanphier, op. cit.
11. Hansen, *et al.*, *Sample Survey Methods and Theory*, vol. i.
12. Everitt, B., *Graphical Techniques for Multivariate Data*, Heinemann, London, 1978.
13. Ehrenberg, A. S. C., *Data Reduction*, Wiley, London, 1975; and Tukey, J. W., *Exploratory Data Analysis*, Addison-Wesley, 1977.

10

Data Interpretation, Presentation, and Release

I have often had occasion to point out to him how superficial are his own
accounts and to accuse him of pandering to popular taste instead of
confining himself rigidly to facts and figures. 'Try it yourself, Holmes!'
he has retorted, and I am compelled to admit that, having taken my pen
in my hand, I do begin to realise that the matter must be presented in
such a way as may interest the reader.

The Adventure of the Blanched Soldier

10.1 UNDERSTANDING AND INTERPRETING DATA

Chapter 9 dealt with the processing of the data to the point of computer
tabulation: there remains a great deal to do before the data are in a format
for release to the users. Or, rather, most users. With the ease of access to
compatible computers which has come about in recent years, many data
sets are transferred on disc or tape from producer to certain users with no
interpretation, analysis, or collation by the producer. We consider in this
chapter the still common need for the surveyor to interpret his own data
and present them in an easily understandable format to users who do not
intend to carry out their own statistical analyses on the primary data.

It is stating the obvious to point out that the surveyor cannot
communicate the significance of the findings to others if he does not
understand them himself. But, sadly, we increasingly come across instances where the basic tabulations have been misinterpreted by the
surveyor. Either significant findings have gone unnoticed or spurious
relationships are produced and stressed by inappropriate applications of
statistical and econometric methods. Many practising survey statisticians
today, including international consultants, are not able to examine a set of
primary tabulations and detect the story that the data have to tell. Rather
they rely on standard software packages to run correlations, regressions,
analyses of variance, and so on, pick out the relationships with the highest
coefficients, and present their findings accordingly. We assert that the
majority of such analyses are invalid and many of the so-called relationships that emerge are spurious.

The easy availability of sophisticated statistical computer software has
produced two deleterious effects: laziness on the part of the competently
trained surveyor, and ease of access for the untrained who do not
understand the limitations of the data set and the conditions underlying
the various analytical techniques.

It is essential for the surveyor to examine the basic computer tabulations to detect gross errors, to assess the internal consistency of the data, to arrive at initial hypotheses about significant differences and relationships that should be further investigated, and to decide how to extract from the (probably extensive and voluminous) print-outs simple two- and three-way cross-classification tables that will convey the significant features in a way that will create the correct image on the retina of the untrained eye. It is this basic exploratory analysis by the surveyor that is being increasingly neglected.

Gross errors can easily arise from an initial simple coding error when the data set is based on a small sample which, for purposes of tabulation, has been weighted up to estimates for the universe, using large weighting factors. Suppose, in a sample of 500 households out of 250,000, the true average expenditure on a recreational item is 10 monetary units including one household with a true expenditure of 100 units. If, due to operator error, this 100 is entered into the data base as 10,000 the estimate of total expenditure for this category in the total population is trebled. We have experienced such a mistake in a recent farm survey.

Many internal consistency checks can be built into the automatic data validation when creating the computer data base. But only when looking at the data in a cross-variable manner will other inconsistencies emerge. Expenditures by certain groups are significantly above incomes. There is a bunching of age declarations on numbers ending in zero or five. Nobody in the sample admits to an income from producing locally distilled beers or spirits, but many in the sample report considerable expenditures on such items.

It is improper to run a series of regressions of all combinations of variables and select those with the most significant coefficients. Depending on the confidence level chosen, one in ten or one in twenty of such attempts will produce an apparent significant relationship erroneously. The surveyor should examine his tables, derive some hypotheses, and draw a few simple two-way graphs to see what relationships may be emerging (and whether they are linear or not) that are consistent with expectations, or confound such expectations.

This step will lead to the decision on which cross-classified tables should be extracted for presentation in a report and which are of no significance, except possibly as a historical record in the files.

These steps, if undertaken by the surveyor—not anonymously by clerks or a machine—will leave him with a much better 'feel' for the data set and the stories to be extracted from it. He will now be in a much better position to communicate these stories to others.

Reminding the competent surveyor of his responsibilities may have a salutary impact. Restraining those who do not understand the proper use of the techniques available in their computer software is more difficult. It is our wish that all analytical software would require the user to indicate his understanding of each technique before he could apply it. Failing such an ideal solution, we list a few ground-rules that apply to many statistical analyses and urge our readers to seek further detailed guidance if these rules are unfamiliar to them.

(a) Errors in the measurement of the independent variables of a function can completely invalidate the correlation or regression coefficients calculated.

(b) Correlation does not establish causation.

(c) A low correlation coefficient does not preclude the existence of a close *non-linear* relationship.

(d) Tests of significance and confidence interval estimation usually require that the residual variation about the fitted model is normally distributed.

(e) In many models there is a requirement that the variability of the dependent variable is constant across the range of values of the independent variables.

(f) Correlations between aggregate values of variables often appear despite the lack of correlation between these variables at the level of the individual. Such correlations are referred to as the ecological fallacy.

10.2 THE PRINCIPLES OF DATA PRESENTATION

Data presentation is complicated when there are a range of users to be served. The difficulty of meeting different needs occurs particularly in table content. The general user wants tables and graphs that highlight the main findings in a way that makes them easy to grasp; the entries in the tables, therefore, need to be rounded to two or three significant figures with liberal use of percentages and ratios. The professional or technician wants the detailed numbers so that he can carry out further calculations. The solution is to partition the report: summary tables highlighting the main features of the analysis in the main body of the text, and the detailed tables with fuller data content in an appendix. Users intimately associated with the survey or its findings may eventually need the computer discs or tapes without editing or abridgement; this kind of access needs special arrangements, which are now increasingly being made via national data banks.

The following principles refer to the presentation of summary tables in the main part of the report. The first three are identical with those mentioned in Chapter 9 for the production of computer tabulations, the others apply more in the present context:

(a) layout clear and pleasing to the eye;

(b) understandable by someone not intimately connected with the survey;

(c) disaggregated at an appropriate level;

(d) layout consistent throughout;

(e) appropriate use of averages, measures of dispersion, and rates;

(f) appropriate use of percentages;

(g) sensible grouping and display of 'all other' and 'don't know' categories;

(h) appropriate use of diagrams, graphs, and charts.

The comments on (a) to (c) in Chapter 9 apply here with an additional emphasis on the careful selection of a minimum of figures. The rounding of numbers, the spacing between them, and the positioning of totals are important and make a considerable difference to the reader's ability to assimilate quickly the information that the writer wishes to present.[1] When dealing with large numbers, the rounding can be combined with the omission of zeros, showing the estimate to the nearest hundred, thousand, or whatever order is appropriate. There is no merit in showing figures such as 104,792.42 if the error margin on the estimate is at least 10 per cent. Therefore, 105 should be shown with the unit of thousands specified in the appropriate row or column heading, or even 100 (rounding to the nearest ten thousand).

The rounding of figures should be done at the final draft summary table stage. If the rounding is made too early, subsequent calculations of ratios or percentages may be in error. Although 12,485 and 97,543 may be shown as 10 and 100 (to the nearest five thousand), the first as a percentage of the second should be shown as 13 not 10.

The choice of class boundaries for grouped frequency distributions may be broader for the summary tables as compared to the original tabulations. Care should be taken not to distort the data as a result. The last group, often shown as '*X* and over', should not contain more than a residual few per cent of the observations, for an open-ended category such as this cannot be interpreted easily during further analysis of the data by the user. The phenomenon of heaping on certain discrete values should not be masked by the use of classes that imply a smoother continuous distribution, or a central point different from the point of concentration. (A recent table showed holdings with a range of 0–5 hectares and an average size of 4.99 hectares—although the 4.99 could be a misprint.) The layout of tables should be consistent throughout the report. The user will have difficulties if, for example, the order of regions in the column or row headings varies from one national table to another. If there is some generally accepted order for these regions in administrative business it should usually be followed. Occasionally there may be a good reason for departing from such a fixed order: the point illustrated by the table may be better demonstrated by ranking the regions in ascending order of magnitude with regard to the variable in question, or the value of a ratio may be similar for a group of regions in one part of the country and very different for the remainder. Consistency of presentation will be helpful, too, if the report consists of a national section followed by separate regional chapters; the user will normally find his way about with greater ease if the order of the topics is kept the same in each chapter.

One of the main purposes of statistical methods is to meet the natural desire for summary measures of distributions. The use of the arithmetic mean needs no explanation; but the use of the mean as the only summary statistic may be misleading. Variables with highly skewed distributions are commonly encountered in household surveys: income is an obvious example. Two regions may have the same mean income per household; but, in one, three-quarters of the households have incomes below the average, whereas, in the other, only one-half of the households are so situated. The median may be a more suitable average to use in such cases and/or the lower and upper quartiles. Measures of dispersion should generally be given—the standard deviation, the range, and, in a few cases, measures of skewness, are the most commonly used. One useful technique for the representation of inequality is the Lorenz curve and its associated Gini coefficient.[2] For example, in some industries a small percentage of enterprises are responsible for a large proportion of production. It is possible to use cut-off points—the proportion of production for which, say, the largest ten companies are responsible; but the Lorenz curve provides fuller information.

Some rates, such as crude birth (death) rates, are justifiably described as crude since they reflect the combined effect of two factors, fertility (mortality) and age structure, and not the single factor of birth (death). Similarly, nutritional measures on a per capita basis may be misleading owing to differences in the age/sex structure of the underlying populations. The difficulties here may be partly overcome by using factors to convert numbers of individuals to numbers of adult equivalents. (Standard conversion tables are available.)[3]

No part can be more than 100 per cent or less than 0 per cent of the total to which it belongs. A newly introduced cash crop may rapidly increase the proportion of total income it provides, rising, say, from 0 to 5, 10, and to 20 per cent; but it is unlikely to maintain this rate of increase, and of course any projection that forecasts that it will provide more than 100 per cent of total income at some future time is nonsense. The idea of levelling off towards an asymptote is useful in this respect and models involving exponential curves may be applicable.[4] In these circumstances a study of the logarithms of the observations may be illuminating.

There are two well-known basic ways of looking at distributions and making comparisons: absolute and relative. For example, in region A, cash and imputed income from maize may total £10 million, whereas in region B it is only £3 million: but the £10 million is only 10 per cent of all income in region A, whereas the lower figure of £3 million nevertheless amounts to 35 per cent of all income in region B. Both the absolute and the relative measures are useful; they indicate different aspects of the same numerical fact. Similarly, income may go up from £100 million to £110 million in a year in region A, and from £20 million to £24 million in region C. The absolute rise of £10 million in region A is larger than the rise of £4 million in region C, but this smaller figure represents a rise of 20 per cent in region C compared with a rise of 10 per cent in region A. Many important descriptive points can be brought out by an appropriate selection and combination of absolute and relative comparisons.

The row or column entries in a table are often also printed as percentages summing to 100 in the 'margins' of the table. Sometimes, but less often, the individual cells of the table are shown as a percentage of the grand total. Most users expect percentages to sum to 100 either in a horizontal or vertical direction and cannot glean much from a two-way set of percentages. When deciding to present either a row *or* column series of percentages a casual choice may be the wrong one. Zeisel offers the following rule: 'the percentages should be computed in the direction of the causal factor'.[5] Causal is used in this context as describing the factor that the surveyor wishes to consider as affecting the percentage distribution which the other factor assumes. Consider a hypothetical set of data presented in a two-way tabulation of income group by size of farm. Table 10.1 shows the income distribution for each farm size. Table 10.2 shows the farm size distribution for each income group.

TABLE 10.1. *Income by Size of Farm (%)*

Income group (£/year)	Size of Farm (hectares)			
	Below 1	1–3	3–5	5 & over
Below 500	40	14	7	6
500–999	30	43	15	8
1,000–1,999	20	29	44	28
2,000 & over	10	14	34	58
TOTAL	100	100	100	100

TABLE 10.2. *Size of Farm by Income (%)*

Income group (£/year)	Size of farm (hectares)				
	Below 1	1–3	3–5	5 & over	Total
Below 500	73	18	6	3	100
500–900	43	43	10	4	100
1,000–1,999	29	29	29	13	100
2,000 & over	18	18	28	36	100

In this example, Table 10.1 brings out clearly the way in which income rises as the size of the farm increases. The picture presented in Table 10.2 is more confused: income is not the cause of farm size, and comparisons using the percentage distributions along that dimension are affected by the very different numbers in the various farm size categories. (In the numerical data on which these tables are based three times as many farms lie below 1 as above 5.)

Causal direction may be indeterminate, ambiguous, or nonexistent, so the choice is not usually so clear; the percentages could run in either direction, each telling a different story. The surveyor must decide which pattern he is interested in, and avoid telling one story in the text when the 'picture', that is the table, is showing the other. Especial care should be taken with regional or other geographic area analyses: sometimes only one direction of percentaging will be of interest, sometimes both may be required to extract the main items of interest. When a geographical region is tabulated against a socio-economic variable, a first approach can be made by taking the region as the 'cause' affecting the variable, with percentages summing to 100 for each region. This rule of thumb does not always work, but the surveyor should be sure that reversing the direction is more suitable for demonstrating his point.

Many tables have a final category of 'all others', 'not stated', or 'don't know'. We have commented in Chapter 9 on the relative merits of a miscellaneous 'all other' category in the context of regions. In the context of answers to specific questions several points concerning these residual categories apply:

(a) 'not stated' is a distinctly different category from 'don't know' and should be treated as such.

(b) the proportion of the total in these categories should be small, otherwise the whole purpose of the table can be questioned.

(c) 'don't know' may be due to a variety of different causes, for example, no knowledge, question was inapplicable, question was badly phrased; the required treatment will vary according to the cause.

Maps, charts, and diagrams should be used to aid understanding and to bring out forcibly the main findings of the report. But they should not be overdone: they will not make their proper impact if there are too many of them. Similarly, any graphs used should not contain many different lines, criss-crossing each other and presenting a basketwork appearance to the viewer. Considerable care should be given to the selection of the variables graphed and the form in which they are presented—original units, percentages, ratios, or other transformations. Experience shows that we must still state the obvious: the axes should be labelled, the units indicated, breaks in the scale (particularly those involving the origin) marked. Remember that if two-dimensional figures are used to indicate magnitudes it is the *areas* that have to be in the appropriate proportion. If photographs can be included they may help the reader to appreciate some of the difficulties confronting the field staff, and indicate the range of communities being studied.

In a number of two-way tables, the diagonal cells are particularly important. Thus the size of the entries in diagonal cells in Table 10.1 highlights the relationship between income and farm size. When a precise relationship is postulated, the relative numbers of cases in the diagonal and off-diagonal cells may provide a useful and quick indication of the strength of the relationship. Similarly, if a graphical representation of the relationship between two variables is required, showing those units which are above or below average, it may be useful to draw a line on the graph representing the average relationship, and the plot of points may illustrate the groupings required for analysis.

10.3 SURVEY REPORT WRITING: THE MAIN REPORT

The required end product of most surveys is a report or reports, the preparation of which is seldom an easy matter. However, if the survey and data processing procedures recommended in this book have been followed, the basic material will be ready to hand. One single survey can give rise to periodic reports to management, a preliminary report to the responsible authorizing agency, a main report, special reports written for technical users, a brief for a minister, a paper to an interdisciplinary conference, a paper for a professional journal, and press releases or a newspaper article. The vocabulary to be used, the setting out of the argument, and the extent of technical and tabular material will have to be adapted in each case to the purpose of the publication. However, there are some general rules of practice that can be regarded as 'good behaviour', which form the basis of a code of conduct. Readers who have read this far will judge how successfully we have followed them ourselves.

Although the objectivity of reports imposes a degree of impersonality in presentation, there is no reason why individual characteristics of personal style should not be retained; they will be beneficial if they make the discussion more forceful, but do not distract the reader by eccentricity. One danger, in particular, should be avoided in writing for an international audience: the use of idiomatic expressions, to arouse attention and to fix interest, can cause bewilderment and even offence when they are not fully understood. For example, we have used the expression 'washing one's dirty linen in public' in Section 8.6, but it might have been better to have used a different phrase.

One way of judging whether the writing progresses with a smooth, acceptable style is to read it over to oneself or to someone else; another is to try it out by having it read and criticized. Easy reading requires hard writing for most of us; usually more than one draft will be required before we can be satisfied that we have done as well as we can. If a phrase seems out of place, but resists redrafting, it may not be needed at all (the cell from which the magician Houdini found it hardest to escape was unlocked).

Good style is likely to come from good manners: the background knowledge and needs of the audience should be kept in mind. Jargon must be avoided, though of course technical terms may be essential in writing for a technical audience. For a non-statistical readership the use of a term such as 'analysis of variance' is to be preferred to the use of a technical abbreviation, such as ANOVA. It will often be better to dispense with the technical term entirely, merely giving an example of its meaning in the context. The reference to 'confidence intervals' later in this section gives an illustration where the technical term cannot be avoided.

Unfortunately a number of words in general use have a special meaning in statistics. One confusing usage is that of 'error'; in statistics this does not necessarily mean 'mistake'. Another is 'bias'; it has been written, 'no one popular word that has been given an exact definition in mathematical statistics has created more of a language barrier between the technician and the layman than the word bias'.[6] Most surveys of human populations will involve at least one of the social sciences, and these also tend to use everyday words in a special sense and to lapse into high sounding and unnecessary jargon. Attempts should be made to put things as simply as possible, avoiding the grandiose phrase. The temptation to 'blind with science' should be resisted: it may succeed in the short term, but in the long term it brings disrepute and suspicion.

Style is important in order to better communicate information but, of course, substance there must also be. This will come from a sound interpretation of the data as discussed earlier. The text accompanying the tabular presentation of the data has the purpose of facilitating the reader's understanding of what the data indicate and, equally important, what they do not indicate, but, perhaps, merely suggest. The writer has to guide the reader to what he may or may not conclude. There is room for great improvement in the quality of such report writing. The unsure writer merely repeats some of the figures in the table, adding nothing, thereby implying that there is nothing of greater note to say. The writer committed to a particular conclusion, by his selection of the facts to highlight, may fall down on his professional responsibility to be objective. And some are just unable to write.

The ability to take a set of tables and write a clear, brief, statistically sound and literate commentary should feature much more prominently in the training of statisticians and allied professionals. The surveyor who recognizes his own deficiencies in this regard should devote some time to reading published survey reports, not for the substance itself, but to see how well the text presents the findings.

The function of the main report is to record:
(a) the objectives of the survey;
(b) the design of the survey and how it was determined;
(c) how the survey was carried out, including an account of deviations from the plan;
(d) a summary of the data collected;
(e) a discussion of the sampling and non-sampling errors affecting the data;
(f) the analysis of the data, including an examination of the results, against a background of previous knowledge; the indication of new findings; and the drawing of attention to results important for policy making.

The main report, therefore, has to cover many topics: but a greater difficulty is that it has to embody the whole survey in a way that satisfies a wide range of readers. Since, as we stressed at the outset, different types of reader require different types of presentation, the task may appear impossible. How can a single report, of a size that is reasonable to publish, satisfy at the same time the local administrator looking for policy guidance, a technical officer looking for detailed information about his speciality, a statistician in another country wanting to benefit from the experience of the survey, and a worker in an international agency looking for data for a cross-country study? Clearly, there has to be a compromise—a trade-off between different needs.

One solution is to partition the presentation into a general section covering, as far as possible, the needs of those readers who are mainly concerned with the basic results; and a technical section that discusses in more detail the design and execution of the survey, and presents a more detailed and/or more sophisticated analysis of the data. If the inquiry has been on a large scale, the report can be spread over a general volume and one or more specialist volumes: if it is a smaller investigation, the more statistically technical material can be put into an appendix, or a series of appendices. Some of the needs of those commissioning the survey will almost certainly have been met before publication by special papers which, finally, will also form the basis of chapters in the main report. Since publication is an expensive business, it may well be that some of the detailed analyses are never published at all. They can be described and made available on special request (perhaps against payment), provided the survey organization can handle the likely demand. Other special topics can sometimes be covered by articles in the relevant journals.

If the report is divided in this way, the section or volume written for general purposes must include sufficient material to ensure that the reader understands any adjustments that have been used and the qualifications which have to be made in using the results. This is not just a case of including this discussion so that the reader can look at it if he wants. Far too many people go to reports looking for figures to support their preconceived ideas; others are looking for unequivocal answers or simple yes/no responses to the problems with which they are dealing. The writer must try to stop the reader from drawing wrong conclusions by including, even in the general section, a sufficient account of the planning and execution of the survey, and the sampling and non-sampling errors attaching to the data. One cannot, of course, completely prevent misunderstanding and misrepresentations, but one should consciously strive to minimize them.

The surveyor arriving at the stage of preparing the report of the survey for which he has been responsible can be in one of two different states of mind. He may be so involved with the survey, and have so much detail to hand, that he cannot select and organize the findings in a way that effectively brings out the main contribution of the work. A large number of findings appear of almost equal interest to him: indeed, some unusual result in a minor area may divert him from important, but less unexpected, findings in the primary field of inquiry. Further, he has lived with the survey so long that he may take the reader's interest too much for granted, and overlook the need for some basic information or definitions in his account. At the other extreme, long and close involvement may generate fatigue and boredom: by the time the surveyor is writing up the material his interest has slackened, and the report is superficial and weary. This is especially likely to happen when the survey has not been fully successful, or when the results only confirm existing knowledge, or are negative. There is no simple remedy for discouragement, but it should be realized that a report of negative results can also be valuable. The writer has not only to show where action is needed, he has also to help stop ill-judged programmes. As mentioned earlier, even if some parts of the survey are not successful there still may be something to save; in addition, an account of the causes of failure will help others to avoid them in future.

One aspect, over which the surveyor frequently has too little control, is the physical form of the final publication. It is worth taking some care about this. We are all familiar with publications whose art paper and glossy covers camouflage poor material, but that does not mean that the surveyor should not get as good quality paper and production as he can for the report. The type should be clear and plain, margins should be adequate, the tables should be neatly set out in a uniform pattern, the diagrams properly placed and reproduced with unsmudged lines, and the binding should hold together. The range and cost of reproductive processes vary considerably from country to country. The difficulties are obvious: we have had to get out reports with inexperienced typists working in a language with which they were not familiar, with poor quality stencils, an obsolete duplicating machine, and soggy paper. The level of facilities has improved substantially in recent years, particularly in offset printing and in inexpensive binding techniques. Whatever the circumstances, the surveyor will nearly always be able to get better produced copy if he takes some trouble over it, and does not accept resignedly what is first offered. Many people find survey reports difficult reading, and a clear physical layout helps them to extract what they want. We have already commented on the number of surveys that have not been written up at all; a number of others failed to make the impact they merited, because the report was inadequately prepared and produced. One effective method is to preface the report with a very brief summary of its major findings. Many potential users will not delve deeply into a large volume unless their interest is attracted by such a summary.

The topics to be dealt with in the main report, set out as items (a) to (f) above, can be related easily to the topics discussed in the rest of the book. The points that should be covered emerge naturally from the discussions and thinking that should have been behind the survey, and the decisions taken about procedures and analysis. Because of this, we make only brief comments at this stage.

After the summary of findings recommended above, the report should open with a section dealing with the design and execution of the survey; this should include:

(a) the objectives and the organization for which it was carried out;

(b) summary of the main issues discussed in the surveyor–user dialogue;

(c) the history of investigations into the topics studied, the information available at the planning stage, and the place of the survey in the exploration of the topics;

(d) the different frames considered and the one adopted (with reasons);

(e) details of the sample design and selection of respondents;

(f) the design of the questionnaire, with copies in the Appendix;

(g) selection and training of the survey team (including numbers in each category);

(h) quality control measures, frame deficiencies, and the extent of non-response;

(i) a timetable setting out the main chronological stages;

(j) the budget and resources employed;

(k) the experiences in the execution of the survey, including time distribution (travelling–interviewing, main survey–post-enumeration, etc.), unexpected difficulties (climatic, political, local hostility).

The main findings of the report should then follow with a summary of the data, presented according to the guidelines in Section 10.2. This section should include a discussion of data reliability and attempt to prevent unwarranted conclusions. This is supplemented by a more detailed and technical section as suggested earlier.

A decision on how to present sampling errors depends upon, and also influences, the decisions about what sampling errors to calculate at the processing stage. Clearly, we are unlikely to calculate sampling errors for all the items estimated, still less to publish on this scale. Here, the US Bureau of Census recommendations are:

When a new, or largely new, sample is selected, sampling variances for a sufficient number of items should be estimated directly from the observed data to provide an adequate base for approximating the sampling errors of all the estimates to be published. Sampling errors should be computed for an adequate number of items from each group of items likely to exhibit different sampling behavior, e.g. groups which reflect different degrees of clustering or different types of crops in an agricultural survey. For single time surveys, sampling errors should be calculated for a sufficient number of items to give an adequate idea of the reliability of the survey estimates.[7]

The sampling errors to be presented to the reader will be the more important ones obtained during the processing; by more important we mean those affecting the major magnitudes dealt with in the survey, and those affecting the survey findings which have policy implications.

Sampling errors can be presented in absolute or relative form, depending on which appears more apt for the particular purpose, but it must be made clear which form is being used. Since confidence intervals of different size (different multiples of the sampling error) may be required for different purposes, it is simplest to quote one standard error, and leave the user to take whatever multiple he wishes. It is sometimes practicable to provide a guide relating the standard error to the size of some selected estimates, and the reader can use the table to approximate the error for estimates whose size is not separately specified in the list. An excerpt from an example given on p. 13 of the Bureau guide is shown in Table 10.3.

TABLE 10.3. *Example of the Presentation of Absolute and Relative Standard Errors*

Size of Estimate	Standard Error	
	Absolute	Relative (%)
25,000	700	3.0
50,000	1,100	2.1
100,000	1,500	1.5
250,000	2,300	0.9
500,000	3,300	0.7
1,000,000	4,700	0.5

The standard error for 66,000 is approximately $1,100 + (1,500 - 1,100) \times \dfrac{(66,000 - 50,000)}{(100,000 - 50,000)}$
$= 1,228$ which we should round to 1,200.

Sampling errors cannot be neglected; but, as we have emphasized, non-sampling errors may well make the larger contribution to mean square error (see Chapter 4). Unfortunately, the non-sampling error cannot usually be assessed accurately, and its extent cannot be quantified in the same way as that of the sampling error. It must be discussed, however, alongside the treatment of sampling error, so that the reader realizes that both types of error exist and what their likely relative magnitudes are. The actual description of how non-sampling errors have arisen and their possible size, will be given mainly in the description of the execution and analysis of the survey (see earlier sections of this chapter). An attempt to estimate their order of magnitude and to combine these estimates with the sampling errors to produce figures of total error should be made in the section on errors. It may well be that little more than a rough grading of quality can be provided, for example:

(a) Basic data are reliable (accurate objective methods used): error is mainly due to sampling.

(b) Parts of these estimates rely on memory recall of the respondent: non-sampling error may be as important as sampling error.

(c) Respondents were reluctant to answer the question on which this item depends: non-sampling error is substantially greater than the sampling error.

In 1981, a proposal for a continuation of a major series of surveys aimed at evaluation of a national development programme included a recommendation to double the sample size in order to bring down the sampling error at the provincial level from 4–6 per cent to 3–4 per cent. We suggested that this might bring about a rise in the non-sampling errors due to poor enumeration and less supervision and that this would more than cancel the gain in precision that was sought. A rebuttal by those overseeing this survey programme included the remarkable statement that there were no non-sampling errors due to the quality of the survey operations. It gave us no pleasure to watch the whole survey programme founder on account of very high enumerator biases that emerged as analyses of the data sets were attempted. Surveyors should deceive neither themselves nor their clients.

10.4 PRELIMINARY REPORTS

The production of a preliminary report may have been foreseen at the planning stage because it was known that the full analysis would take some time and orders of magnitude are required quickly; or because partial results were needed for urgent policy requirements. In these circumstances, an analysis of a sample of the survey returns should have been provided for in the analysis programme. There are considerable dangers if the preliminary report is based on only the early returns received, since these may not represent properly the whole range of the data. The earliest reports may come from the areas easiest to survey, or from the areas nearest the regional offices. Further, although any report can be misunderstood or misreported (consciously or unconsciously), a preliminary report may be more easily misconstrued.

There will be occasions when there is a requirement for a preliminary report that was not foreseen when the survey programme was planned. A development programme may be under consideration, a project in one region is being prepared, or there is a sudden need for background information on household structures against which to set results of an educational or health report. A surveyor should be prepared to defend his timetable to meet the responsibility of preparing the main survey report. But, within the context of an official survey, it is necessary to be responsive to national needs. There are two options that should be given serious consideration. First, it may be possible to alter the order of the data processing and analysis of survey data to give priority to the regions or other administrative areas of interest. A small disruption caused by such a change is a small price to pay for greater user satisfaction, which may translate to greater user support for later surveys. Second, a subsample of the survey returns can be duplicated and analysed by staff concerned with the programme for which the data are required, thus causing no serious disruption to the original survey timetable. These unanticipated users may be in a position to finance such a parallel analysis and even make a contribution to facilitate the main analysis.

The recommendation to be more responsive to urgent requests is more forceful than that in the first edition of this book. Experience in recent years has indicated that official statisticians and surveyors are becoming too defensive and isolated. A more flexible and outgoing attitude is now required.

10.5 FURTHER ASPECTS OF THE PRESENTATION AND RELEASE OF DATA

Most surveyors would like to make the results of their survey available to those who have given their time to answer questionnaires and to provide other information. It is sometimes relatively easy to do this: the publication of an industrial production census report at a reasonable price accompanied by a press release will meet the needs of most of the enterprises that provided the information. There is no similar easy way to make any direct return to the population as a whole. The most useful action the surveyor can take is to do his best to ensure that his findings are properly employed in policy making. This is a difficult task in which his technical training may be of little direct value.

Administrators have their own set of priorities: they usually have a shorter time scale on which to operate than surveyors, and often have to take decisions before full and reliable information is available. Their impatience with surveyors frequently adds to the difficulties that the latter already face because of their relative lack of influence in the government machine. One part of the way to deal with this problem comes when the survey is planned, as the objectives of the survey are crystallized in the user–surveyor dialogue, and the final plan including the timetable and analysis is agreed. Another part comes by carrying out an efficient survey and so fulfilling the 'contract'. The final episode is to take especial care about briefing the users on the findings of the survey. They will probably react most to the preliminary report: they will no doubt want to see the main report, but by the time that it is available they may have made their decisions.

In accordance with the previous discussion, success will usually follow if the needs of the recipient have been adequately gauged. The material included in a brief, or a series of briefs, will have to be tightly selected, results set out within a framework of general policy, and relevant findings (not always statistical) in the problem area will have to be displayed. Technical explanations should be cut to the minimum; and relatively little tabular material should be given in the main body of the draft, although more can be included as an appendix. The whole thrust of the argument will be closely directed to specific issues. The approach must be different from that in an account for general purposes, such as that given in the main report. This does not mean that there is any slackening of attention to the adequacy of the evidence; it is even more important. Paradoxically, the greater economy of style means *less* can be left to the reader. The surveyor has to show he can meet a main requirement of the administrator—for example, market information, or an early estimate of harvest size. In certain cases the first job is to show that there is a statistical gap to be filled, and that it is in the administrator's interest to provide the resources to have it filled.

A surveyor has to be more than a good professional if he is to ensure that the results of survey work play their proper role in policy formation: entrepreneurial ability is called for. Not all surveyors find this to their taste, and these may develop into 'back-room boys'. They will provide drafts, but their work needs some editing to make its full impact. There is nothing wrong about this kind of division of labour, provided the resources are sufficient to allow it.

Much of the discussion in this chapter has been general and the reader will need to supplement it by the study of a number of survey reports. It would be invidious to mention 'good' or 'bad' surveys, and probably pointless in any case, since many of them circulate primarily in their country of origin. Any practising surveyor will inevitably acquire a number of reports and we hope he will be able to assess them more effectively after reading this chapter, and so develop his own style.

Notes

1. Ehrenberg, A. S. C., *Data Reduction*; and 'Rudiments of Numeracy', *JRSS* (A) 140, 1977.

2. Discussed in a number of texts: see, for example, Fuller, M. F., and Lury, D. A., *A Statistical Work Book for Social Science Students*, Philip Allan, Oxford, 1977, Chapter 5.

3. Joint FAO/WHO Expert Committee Report, *Energy and Protein Requirements* WHO, Geneva, 1973; Feigehen, G., and Lansley, P. S., 'The Measurement of Poverty: A Note on Household Size and Income Units', *JRSS* (A) 139, 1976, pp. 508–18.

4. Yeomans, K. A., *Applied Statistics*, Penguin, Harmondsworth, 1968. For a basically graphical approach, with practical examples, the Gregg, J. V., *et al.*, ICI Monograph No. 1, *Mathematical Trend Curves: An Aid to Forecasting*, Oliver & Boyd, Edinburgh, 1964, is still useful.

5. Zeisel, H., *Say It with Figures*, Routledge & Kegan Paul, London, 1958, p. 24.

6. Hansen *et al.*, *Sample Survey Methods and Theory*, vol. I, p. 16.

7. 'Standards for Discussion and Presentation of Errors in Data', Technical Paper 32, Bureau of the Census, US Department of Commerce, 1974: rearranged and reprinted as No. 351, Part II of JASA, 1975, p. 10. Another treatment, and a discussion of the entire subject of this chapter, is given in UN, *Recommendations for the Preparation of Sample Survey Reports* (Provisional Issue), Statistical Papers, Series C No. 1, 1964.

11

General Household Surveys

'Your own little income,' he asked 'does it come out of the business?' 'Oh no, Sir, it is quite separate and was left to me by my Uncle Ned . . . as long as I live at home I don't wish to be a burden to them, and so they have the use of the money just while I am staying with them'.

A Case of Identity

11.1 THE IMPORTANCE OF THE HOUSEHOLD FOR SURVEYS

References to households as sample units, and to heads of the household as respondents, recur throughout this book. One reason for this popularity of the household has already been mentioned—the suitability and relative availability of household lists, derived either from a recent census or a special listing within a sample cluster, for use as sample frames. This convenience would not by itself lead to the pre-eminence of the household. The decisive factor is that, in developing countries, the household is the most common unit of production as well as of consumption.

The majority of the population living in rural areas and dependent mainly on agriculture cultivate, harvest, sell, and consume agricultural commodities that the household has produced as an operating unit. The household is bound together not merely by social ties, but by economic forces: it still pays, in survival terms, for the household to operate as a single unit. This may not be true for certain groups, such as pastoralists, where cattle are managed by clans or extended groups. Further, as urbanization proceeds, the household as an economic production unit starts to lose its coherence, and it may well be that land shortage following rapid growth of population will, finally, also weaken it. The absolute number of agricultural workers in rural households is expected to grow for some years yet, although the proportion they bear to the total economically active population is already falling in many countries.[1] Thus, while the usefulness of the household sample unit will probably decline as the population moves into towns and becomes reliant on wages or the informal urban sector, at present—and for some time to come—the household is both the most readily available and the most relevant unit for a wide range of survey purposes. The increasing attention being paid recently to levels of living and to income distribution reinforces its position.

A proposal for a Household Survey Capability Programme, prepared by the UN Statistical Office, has this to say about the role of household surveys:

There is universal recognition of the key role of the household sector in the socio-economic development of developing countries. Households account for much of the productive activity and are themselves affected by economic and social changes. . . . A continuing programme of household surveys generates integrated data on a wide range of subjects—income, consumption and expenditure, labour force and employment, housing, water supply, health, nutrition, education and activities of household enterprises. . . .[2]

A survey with the household as the sample unit does not necessarily mean that the survey is strictly a household survey. The proposal quoted above goes on to mention fertility as a common subject for a household survey. But most fertility surveys are concerned with a sample of adult women, and fertility rates are expressed in terms of women not households. This type of survey uses the household as a sample unit, but covers the subset of the population eligible for the survey within the household. We prefer to call a household survey one in which the household is the final sample unit and also provides the unit of reference for the major focus of the survey. Questions may be asked of individual members of the household, but these are aimed at itemizing individual components of the household activities in order to obtain the household data by aggregation. Surveys of subjects such as fertility, child nutrition, food intake, and energy expenditure are surveys of *individuals*, although the individuals are also household members; and membership of the household will determine, to some extent, the individual's fertility, health, and consumption habits.

There are topics of inquiry where the household is not the best sample unit. The data refer to individuals, are to be analysed in that form without reference to the household of which they are members, and the individuals themselves are a particular subset of the population. The choice whether to use a household as a sample unit in surveys essentially dealing with individuals is considered in the next chapter.

11.2 DEFINITION OF A HOUSEHOLD AND ITS HEAD

Although there is general agreement that the household is both a convenient and appropriate sample unit for many surveys, there is no universal agreement as to its definition. Not only are different definitions adopted in different countries, but within one country the definition may change from one survey to another, resulting in incompatability and consequent loss of analytical comparisons over time or between groups. The stipulation that a definition must be adapted to local conditions has been stated as follows:

a precise definition can be developed only by considering the social structure and type of living arrangements found in the survey area. . . . Families that are related to each other may have close economic ties without living under one roof or eating together . . . husbands and wives may maintain separate dwellings and/or separate finances. . . . Consultation with knowledgeable persons, particularly anthropologists, should be helpful in arriving at a definition of household appropriate to the area in question as well as the research objectives.[3]

The authors of this quotation may, in their global context, be referring to 'area' as a country; but the implication could be that the definition must be varied and adapted to local conditions within countries; although this is also a sensible course, the variations which can be allowed within one country must be limited if the data from successive surveys are to be closely related.

What further guidance can be given? A UN Manual recommends:

The concept of 'household' is based on the arrangements made by persons, individually or in groups, for providing themselves with food or other essentials for living. A household may be either (a) a one-person household, that is, a person who makes provision for his own food or other essentials for living without combining with any other person to form part of a multi-person household or (b) a multi-person household, that is, a group of two or more persons who make common provision for food or other essentials for living. The persons in the group may pool their incomes and have a common budget to a greater or lesser extent; they may be related or unrelated persons or a combination of both.[4]

A search for simplicity prompted the authors of the Manual to supply 'a more practical form for everyday use by the enumerator', namely that 'a household is a group of people who live and eat together'.

The attempt to explain to the enumerator what is meant by 'living and eating together' leads to variations in the operational definitions adopted. A few examples will illustrate the point.

(a) A survey in Jamaica used the following definition:

A group of people who live together and draw from a common or pooled fund for all their major items of expense . . . , a servant or other home help who sleeps on the premises (even in a separate building) and shares meals . . . is accounted a member of the household.[5]

(b) A budget survey in the neighbouring Bahamas defined a household as:

those persons living together under one roof and acting as a single unit for budgeting and consumption purposes and for purposes of providing themselves with food and other domestic arrangements.[6]

(c) In Kenya, several variations have been tried. The 1974 survey used:

a person or group of persons normally living together under one roof or several roofs within the same compound or homestead area and sharing a community of life by their dependence on a common holding as a source of income and food, which normally, but not necessarily, involves them in eating from a 'common pot'.[7]

(d) The Philippines adopted a shorter form:

a group of persons who live together under the same roof and share in common the household food.[8]

These examples highlight the three features that surveyors rely on in deciding household composition, namely:
(a) a common source of the major part of income;
(b) sleeping under one roof or within one compound;
(c) a common source of food.

The difficulty is that the three items do not always all apply to the same group, and they may conflict with each other. None of the three items is always easy to establish. So phrases such as 'normally but not necessarily' creep in, or only a simple definition, as in the Philippines, is retained with no attempt at an all-encompassing cover.

A group comprising a husband, several wives, and their children may be considered as one household. Yet each wife (with her children) may live in a separate hut, cultivate her own piece of land, and prepare meals for herself and her children independently of the other wives. To separate the wives each into her own household comprising a wife and her children leaves the male to be allocated arbitrarily to one or to be shown as a single-person household, unless there is some special position of first wife. This is just one example of an exception that, depending on the definition adopted, may make a real difference to the distribution of households, since the average household size and the percentage of female heads of household will change substantially depending on the definition used. We know of successive surveys where slight changes in definition reduced the mean household size by one, and others where the proportion of female household heads was halved.

The problem is that the selected tangible features are an attempt to define an intangible quality of 'community of life' that can be recognized, but not easily defined. We venture the following definition of a household as a starting point:

A household comprises a person, or group of persons, generally bound by ties of kinship, who live together under a single roof or within a single compound, and who share a community of life in that they are answerable to the same head and share a common source of food.

The ties of kinship have to be specified within the particular cultural context. The sharing of a common source of food is taken to include the provision for a common income source; in rural areas much of the food will be produced by the household and in an urban or non-agricultural environment a considerable proportion of income is used to purchase the food.

We stress, however, with the UN Manual quoted earlier, that: 'More important than defining the term is to list the specific border-line cases which cause uncertainty.' Is the household definition to include temporary visitors, or is there to be a residential qualification (see comments on *de facto* and *de jure* population below)? Are boarding guests, lodgers, or servants to be included in the household? Should a group of unmarried persons sharing a lodging be regarded as a household? The most appropriate answers to these questions, and others, will sometimes differ according to the purpose of the survey; but special needs have to be offset against the desirability of following a standard national practice, if there is one. Decisions should be taken and carefully taught to the enumerators, since in practice the worst confusion is caused by enumerators' adopting different procedures for the inevitable cases they cannot fit into whatever definition is adopted.

If the sample has been drawn from a previously established list of households, the definition used in preparing the list is, to that extent, imposed on the surveyor. To split or aggregate households in the sample at this stage may cause problems in the weighting and analysis of the data.

Once a household is identified, a decision must be made whether to enumerate the *de facto* or *de jure* population, or both. The *de facto* population is that present at the time of enumeration, whereas the *de jure* population includes those who are normally resident in the household even if they are absent at the time of enumeration (and excludes those who are present but not normally resident). In a sample survey conducted over a period of time the *de facto* population approach is not very convenient; since it is based on presence in the household on a particular night it will be subject to fluctuation from visit to visit. The *de jure* population is usually more relevant for purposes of multi-visit surveys, and so 'normally resident' must be defined. The definition should include a qualifying period. How long has this person lived in this household? Or, if the person is temporarily absent, how long has he been away and when is he likely to be back? A United Nations recommendation is that six months of continuous residence in the previous year is required to qualify as 'normally resident', but the UN Manual on Demographic Surveys describes complications that can arise when using such a definition and suggests a question such as, 'In what dwelling have you lived most (or longest) during the last twelve months?' Such a pragmatic resolution of the problem is attractive, dealing, as it does, with cases of more than one move in the last six months and cases of dual residences, for example, an urban worker who returns to his rural house at weekends. It still remains open to difficulties of referring to 'a year'.

In the event, the enumerator may find his judgement influenced by the feelings of the respondents. If household members themselves say that someone is regarded by the household as resident, it will be difficult to omit the person from the list. There is much to be said for accepting the concept as it is applied by the population. Any bias that is introduced is likely to be less than that caused by serious enumerator–respondent misunderstanding if arbitrary and alien concepts are insisted upon. The corollary is, however, that an attempt to find out and record this local interpretation of residence has to be made, and its implications reconciled with the definition of household in use. Needless to say, the best approach is to have a pilot inquiry after which a satisfactory definition of residence can be determined, which can be uniformly applied by all enumerators during the survey proper.

The attitude of the potential respondents has relevance, also, in the other definition problem encountered in household surveys: namely, who is the head of the household? In Chapter 13, which deals with holdings and the holder, attention is given to the definition of the holder, as serious biases may be introduced if that definition is not carefully applied. Nevertheless, in the case of the household head, the definition, however worded, describing the operational, economic, and social responsibility vested in the head is likely to be subservient, in practice, to the opinion of the household. To quote the UN Manual once more: 'The best procedure is simply to accept as head the person who is stated to be head by members of the household; there will rarely or never be any disagreement.' The head of the household may be absent at the time of the survey, but is regarded as normally resident. The case of an absent male described as head of the household by the respondent (perhaps his wife) who, on further enquiry, turns out to be spending most of his time in another household in which he would also be regarded as the head, is more puzzling: for example, an urban worker with a wife in town and a wife resident at his rural home. In such a case, when the putative head cannot be regarded as resident, he can scarcely be recorded as head of the household without introducing an element of double counting. If a decision to avoid double counting is necessary (as it frequently is), the best procedure may be to record him in the field but delete him at the analysis stage. The correct person to interview in a household survey is the head of the household or a person designated by the head, perhaps his wife in the case of an absent working male. (See Chapter 8.7 for further discussion on this point.)

11.3 MULTI-SUBJECT HOUSEHOLD SURVEYS

In Chapter 4 we included within the classification of surveys the multiple (integrated) survey and the multiple (successive) survey. Household surveys are rarely limited to one narrow subject; even if the prime objective is to measure income the questionnaire will inevitably cover other socio-demographic ground if only to provide necessary explanatory information against which the income data can be reviewed. But how convenient it would be to cover a wide range of subjects in one survey round. Why should there be a limit to the number of subjects included in an integrated questionnaire? Clearly, there has to be some limit imposed by the length of interview or number of closely spaced successive interviews a household can be subjected to. And, as we stated in Chapter 4, many would argue that the subjects treated in a multiple subject survey should be closely related, for example, income, expenditure, production, and consumption.

Multi-subject surveys, as recommended by international agencies, have been based on a successive programme of surveys over time, following the arguments above. But, recently, a truly integrated multi-subject survey of living standards has been piloted which covers twelve subjects in one survey round.[9] Its authors argue that a wide range of information is required for the same household at the same time period in order to undertake formal multivariate analysis that will contribute to a fuller understanding of levels of living and their determinants. They could claim that the concept of living standards is the common link between the subjects covered, thus justifying the scope even on the grounds advanced above. There are, however, major problems with such an all-embracing approach.

First, the size of the questionnaire becomes daunting—even with tabular layouts used to the maximum extent the questionnaire, in the example quoted, runs to over sixty pages. This translates into long interviews—longer than we regard as desirable on grounds of ethics and quality of response. In consequence there is the danger of superficial coverage by the enumerator.

Second, the subject coverage is so wide that different members of the household need to be interviewed and children measured. There is a confounding of subjects that truly refer to the household as a unit with those we regard as relating to individuals for which there is no household equivalent, such as women's fertility history.

Third, and most important, such wide-ranging integrated multi-subject surveys necessarily assume that one common sample design is appropriate for all subjects, or at least can serve adequately the requirements of all subjects. This is likely not to be the case. We have discussed in Chapter 6 the sampling efficiency cost that is paid when collecting data on some variables using clustered samples—but such a sample design is very efficient for other variables. All subjects must be made to fit within a common interview mode involving recall over extended periods for variables, some of which are not easily reported in this way.

On balance, therefore, we confirm our preference for limiting the integrated multi-subject surveys to those subjects that are clearly directly related, require the same respondent, exhibit the same type of variation and skewness so that a common sample design is appropriate, and which do not require an interview of longer than one hour. Nevertheless, the attraction of attempting to meld into one survey a range of subjects that would otherwise be spaced over several years is not felt only by econometric analysts. Survey offices, limited in resources, whose budgets for recurrent expenditures to manage field operations are not secured for the future, may well wish to seize the opportunity offered by the funding of a household survey in one particular year to cover multiple user needs. Recent experience indicates that, at least, such an effort is not as impracticable as it might seem, and demonstrates that more can be obtained in a first interview than many advocates of objective observations and measurements and multiple visits believe. But there are costs and these must be paid.

11.4 INCOME AND EXPENDITURE SURVEYS

Income and expenditure, or budget, surveys have been popular in the early years of statistical development in many countries. Early examples frequently involved urban workers in order that hourly wage rates could be reviewed. More recently, budget surveys with a national coverage feature high on most countries' survey programmes. In urban areas the concept of income is reasonably unambiguous, but such is not the case in many rural, agricultural areas. Much of the income of an agricultural household derives from sales of produce that take place intermittently. Such households also usually rely on their own produce for part of their food needs, but this production will not be reported as income, even if the respondent knows the amount or its value. Estimation of the value of production raises conceptual and definitional problems that have bedevilled rural income surveys. Net or gross production returns? If the former, net of what, exactly? Purchased input costs certainly, and hired labour; but what of family labour? And in valuing the production, which is the appropriate price, farm-gate, rural market, official marketing agency; and at what season? There is no one correct answer to these questions: we urge the surveyor to be aware of them and adopt a set of definitions, train all concerned in their application, and ensure that they are consistently applied. It is failure to recognize the complexity of an income probe that has led to the common failure to report such information from budget surveys.

The questionnaire, for expenditure at least, is simple enough to design, consisting essentially of lists of items. The data are recorded in value and sometimes quantity terms that leave little room for ambiguity and gross misstatement. An example was shown as Figure 6.5. Features of these surveys that require particular attention include:
(a) a satisfactory level of response is not easily achieved;
(b) audit of returns by supervisors within one day of enumeration;
(c) frequent visits to sample households;

(d) careful timing of visits.

The selection of a sound initial sample of households presents no difficulties, other than those present in any household survey. Stratification by approximate income levels, using a proxy indicator if necessary, will certainly be of advantage. The use of place of residence (that is, stratification by area) may be a useful proxy, but in some urban areas may result in substantial misspecification. However, when the intensity of the survey and the intended detailed probe into all sources of income are explained to the prospective respondents, there is likely to be a high refusal rate, unless great pains are taken to obtain co-operation. Fatigue or annoyance, owing to the repeated visits by the enumerator that are necessary to complete the survey, may cause 'drop-outs' during its progress, presenting not only further non-response but also wasting the resources already devoted to that household.

This type of survey, more than any other, therefore requires a great effort by all concerned in the area of public relations. Selected households must be fully briefed and encouraged to participate in the survey, and must then be 'nursed' through it. Refusals in advance of the survey, and losses during it, may mean that the sample successfully enumerated is biased. It can rarely be assumed with safety that refusals and 'drop-outs' are similar in their income and expenditure patterns to those enumerated. Replacement procedures, even though taken from a pre-selected list in accordance with probability theory, still involve this assumption. Replacements should be limited to genuine non-contacts that are established as resulting from changes in the household population since the frame was drawn up and the sample chosen.

One of the advantages of budget surveys already mentioned, is that the data required include items for which other sources may already exist. The price of a kilo of sugar, the cost of cigarettes (whether in terms of a packet or an individual unit), the prevailing price of a 'heap' of potatoes in the local market, are either known or can be collected. An unskilled labourer's wage is known, within a certain range: so balancing checks between household expenditure and income can be made. The data recorded by the enumerator should be checked by a supervisor against price lists and schedules of prevailing wage rates, etc. In addition, the supervisor should keep a cumulative account ledger for each household in order to maintain a watchful eye on the balance between reported income and levels of expenditure. These checks must be made with a minimum of delay. The discovery of erroneous data after a delay of one week or more may mean that the respondents cannot recall the necessary details, or opt out as a result of the repetition, and the household concerned has to be excluded from the survey. The editing role of the supervisor in the field is more important than that in the office, and in-depth field editing is a prerequisite for success in budget surveys.

The required frequency of visits to each household is a matter that has been debated at length. In Chapter 6 we discussed recall and reference period problems, including the present context of budget surveys. We remind the reader that the total period of the enumeration for any one household should take into account the need for a 'run-in' period when both enumerator and respondent establish mutual confidence and the 'prestige effect' is overcome. In other words, if the survey is planned to cover the period from one pay-day at the end of a month to the next at the end of the succeeding month, it is advisable to commence the survey a few days in advance of pay-day and to discard these early records. Seasonal effects must be taken into account in deciding the total duration of the survey. These effects are particularly important in rural areas where income generation is seasonal. Long survey periods will require the rotation of the sample on, perhaps, a monthly basis.

The timing of the enumerator's visit, in terms of hour of the day, is also an important consideration. From the point of view of the surveyor (but not, perhaps, that of the enumerator), evening is the best time for enumeration, because the reference period can include the day of the enumeration, expenditure having been completed for that day. This ensures that the reference period is at least 'closed' at one end (see Chapter 6). Daytime calls with the reference period closing on some earlier date risk transference of expenditure into or out of the reference period. Whatever the wishes of the surveyor and the convenience of the enumerator, the convenience of the respondent must be paramount. It is bad survey practice to interview junior members of the household during the day in the absence of the head of household and his wife; and it is not appropriate to conduct a household budget interview at the respondent's place of work.

The above review of the demands made by a survey of this type leads to the conclusion that the chances of success are much improved if the survey is limited in geographical spread. A national income and expenditure survey runs into great difficulties with public relations and with the field supervision. If areas are rotated, for example, if a series of sequential inquiries is mounted in the major towns, the design and analysis must provide for the separation of seasonal and area variation. If a continuous household survey capability exists on a national scale, it should, of course, be utilized. But a much larger and stronger force of professionals and experienced field organizers must be provided for this type of survey than for single-visit household surveys.

The following quotations from an unusually frank report of a budget survey conducted in the 1970s provide a perfect example of how badly things can go wrong.

Random number tables were used to select 50 enumeration districts. Once a district had been selected a map of the area was taken and 5 blocks evenly distributed through the area were marked. Each enumerator was asked to select two houses within each block. . . . After a few weeks of the survey it was realised that this method was quite unsuitable. An entirely new approach was used whereby we approached 80 employers . . . and asked each to furnish us with 10 employees who fitted into our target frame and were able and willing to take part.

The pre-test had to be completed within office hours as there was not provision for extra payment to be made to the enumerators. This effectively cut out of the sample any family where there was no responsible adult at home before 5.30 p.m. . . . Our sample was, therefore, restricted to a very small section of the population.

The enumerators soon realised that to question families in a poor looking house in a poor area was a waste of time, so they only approached a stone house, a new house or a house with a car outside, which further biased the sample.

She (the enumerator) also complained that repeated recalls were necessary as many in her area got home late; however, as it grew darker, the area became less safe. Finally, she visited the houses in the morning, but with no more success.

The enumerators found great difficulty in finding any householders willing to take part. . . . The survey questionnaires unfortunately required information on personal and household income and this particular question seemed to spark off a lot of trouble. . . . Just after (the previous budget survey) was completed a Real Property Tax was introduced and many people erroneously connected the two.

It is scarcely surprising that the survey had a non-response figure of 418 households out of 560 chosen!

The sensitivity of income questions is well known, and often increases as incomes get higher. If the visits are well timed and frequent, a high level of accuracy in the recording of expenditures may be obtained, but ways of ensuring an accurate statement of income are difficult to find. If the householder is convinced that his answers will be treated confidentially, if he understands and sympathizes with the reasons for the survey, and if he finds the enumerator likeable and trustworthy, his income may be fully revealed. But it is not enough to obtain just the head of household's income. Incomes of other household members and informal incomes from part-time activities need to be carefully and tactfully probed. Income calculation is complicated further by remittances, loans, and gifts in cash and kind. Details on these inter- and intrafamily transactions are difficult for even co-operative respondents to elucidate. If, despite the best endeavours of the survey team, incomes are seriously underestimated, total expenditure may have to be used as a proxy for income.

11.5   FOOD CONSUMPTION SURVEYS

A household food consumption survey must be distinguished from a dietary or nutritional survey. Indeed, the title 'food consumption survey' is often something of a misnomer. Measurement of actual food eaten is the function of the dietary survey. A food consumption survey, as commonly defined, measures the food available to the household, within the household. Food eaten outside the household is usually included in the questionnaire, but proves very difficult to measure in the context of a household inquiry.

It is difficult to make a case for conducting surveys that have the measurement of food consumption (food 'expenditure' as it were) as their only objective. More logically, such an inquiry forms a part, often a major part, of a household budget survey. Oddly, although many budget surveys do collect data on the quantities of food entering the household, these data are not used in this form, and the quantities are merely converted to values in order to compute total expenditure. There is very little that has to be included in a food consumption survey that is not found in many budget surveys; more detail in the itemization of food may be all that is required. A programme for food consumption surveys, prepared by the FAO, has this to say on the subject of integration:

Bearing in mind considerations of cost, purpose and expediency, it may be found that the choice is not so much between a separate household food consumption survey and one which forms part of a household budget survey, but between a household food consumption survey to be carried out for only a limited period and one which will be permanently incorporated in a household budget survey to be carried out continuously or at regular intervals. Faced with such a choice, the advantage appears 'prima facie' to be overwhelmingly in favour of the latter.[10]

The programme goes on to consider modifications that may be needed because of the dangers arising from an overloading of the budget survey, and reviews the arguments in favour of independent food consumption surveys. If, for example, the variation in food consumption from week to week is much less than the variation in expenditure, 'resources might be better devoted to obtaining a larger number of food records over a shorter period of reporting than that usually adopted for a budget survey'. This brings us back once again to the thorny issue of recall and reference periods.

Overloading may be a problem when various surveys with widely differing objectives are put together in the framework of a multi-subject household survey as discussed earlier, but the general merging of household budget and consumption surveys seems a natural and logical procedure, and should be fully explored. The main issue when considering food consumption is, therefore, the measurement of quantities rather than values. There is, of course, no difference for household production consumed at home as quantities have to be measured in a budget survey as well, in order that values may be imputed. Similarly, quantities have to be measured to estimate the value of gifts and transfers in kind.

Quantities can be estimated through questioning at interview or by physical measurment of weight, volume, or number. There are two levels at which each of these techniques can be applied: they can refer to the quantity at the time of purchase or harvest for domestic consumption, or to the quantity immediately prior to preparation for consumption.

If the quantities refer to the amount 'entering' the kitchen store, then it is necessary to record the period over which this quantity was consumed. For example, sweet potatoes may be harvested in just sufficient quantity for one or two meals, whereas one bag of maize may be kept for gradual domestic use over a period of one or two months. Items that are stored in bulk will therefore have to be inventoried at the start and close of the survey of each household. A distinction should be made between the farm store and the kitchen store, if, indeed, these are differentiated by the household. We are not concerned in a food consumption survey with the farm store but with the stock from which the housewife draws directly for the preparation of meals.

If instead of monitoring depletion of the kitchen store, quantities used for each meal are estimated, the demands on enumerator and respondent are greater. Each meal must be dealt with in turn. This approach is the only option for food items purchased or harvested direct for the 'pot', that is, items that never enter the store. Direct weighing of food as placed in the store may be practicable, but weighing of the ingredients for each meal is not, unless the enumerator is maintaining almost continuous contact. So, in a survey with a large number of respondents, where such continuous observation is not possible, the quantities used at each meal must be obtained by interview, with the use of appropriate physical or visual aids.

There are often advantages in a mixture of the methods. Food brought into the household (whether purchased or harvested) should be recorded and weighed where possible. Quantities used each day should be estimated by interview, using a short recall period. Biases in this reporting should be estimated by a periodic measurement of the quantities remaining in the store.

One difficult problem remains. How is the quantity of food consumed outside the home to be estimated? In the context of populations discussed in this book, consumption outside the home is not mainly in terms of restaurant meals; rather, it is consumption of items such as sugar-cane picked and eaten in the field, snacks purchased from wayside vendors, and food consumed at weddings, funerals, or other special events. Approximate information can be obtained by interview. The interviews, however, need to be conducted with each household member: this adds to the burden of the survey, and, after this effort has been made, often turns out to add little to the aggregate consumption figures. This part of the inquiry may be restricted to a subsample of households.

We repeat that a food consumption survey is not usually designed to produce estimates of food intake but only levels of food utilization. With this in mind, the extent of guest participation in household meals and gifts of food by the household to non-members of the household can be recorded in terms of frequency of occurrence, but complicated procedures for estimating the precise amounts involved may not be necessary.

The difficulties outlined above explain why so many food consumption surveys produce what appear to be, from other information, sizeable overestimates for substantial numbers of the households included. It is not just that average figures seem high, the distribution of food consumption in income and calorie terms often shows a high proportion of poor families disposing of food supplies well in excess of 3,000 calories per consumer unit per day.[11] Resulting estimates of malnutrition would be lower than other evidence suggests. Despite this danger of overestimation, useful data on consumption patterns and distributions may nonetheless be obtained, and may be worthwhile, particularly as part of a budget survey.

Surveys which seek to measure the actual intake of food by household members face the most severe difficulties in measurement since the accuracy required can seldom be obtained without disrupting and distorting the household way of life. These surveys are best left to specialists on a case-study basis. The data are often required to provide answers to specific questions related to a limited geographic area, and linked to food supply or other direct intervention programmes aimed at vulnerable groups in the population. As measurement of the quantity of food actually consumed must refer to specific individuals we discuss dietary surveys further in the next chapter.

Notes

1. FAO, *Agricultural Employment in Developing Countries*, Rome, 1973.
2. National Household Survey Capability Programme: Draft Outline of the Proposal. Tabled at UN Statistical Commission, New York, Feb. 1979.
3. Freedman, D., and Mueller, E., 'The Standard Package: The Household Roster' in Brown, J., et al. (eds.), *Multi-purpose Household Surveys in Developing Countries*, OECD, Paris, 1978, p. 229.
4. *Manual on Demographic Surveys in Africa*, UNECA/UNESCO, Sept. 1974.
5. *Expenditure Patterns of Working Class Households, 1963–64*, Dept. of Statistics, Kingston, Jamaica.
6. *Household Budgetary Survey Report 1970*, Commonwealth of the Bahamas, Dept. of Statistics.
7. *Integrated Rural Survey 1974–75*, Central Bureau of Statistics, Ministry of Finance and Planning, Republic of Kenya, March 1977.
8. Mijares, T., 'Household Surveys in the Philippines' in Brown J., et al. (eds.), *Multipurpose Household Surveys in Developing Countries*, p. 131.
9. Ainsworth and Muñoz, *The Côte d.Ivoire Living Standards Survey*.
10. Clayton, S., *Draft Programme and Guidelines for Food Consumption Surveys*, FAO, Rome, 1978.
11. *Fourth World Food Survey*, FAO, Rome, 1977.

12

Surveys of Individuals

'She has at least an answer for everything.'

The Naval Treaty

12.1 RETAINING THE HOUSEHOLD AS THE SAMPLE UNIT

For many surveys, as we have seen in the previous chapter, the household is the logical unit for reporting data relating to socio-economic variables. Other surveys use the household as either the penultimate or ultimate sampling unit because it is a conveniently identifiable unit, not because the data will be reported in terms of household aggregates or values. In the next chapter we will discuss using the household as an entry point to the identification of a holder and the holding associated with the household. On other occasions the household acts as the location for the identification and selection of individuals to whom the inquiry is addressed. Is the household an appropriate mini-cluster in a sample design for such a survey? In answering the question the following issues must be taken into account:
(a) the number of individuals not living in households;
(b) the definition of the subset of the population of relevance to the study;
(c) the frequency of occurrence and geographical distribution of the main attributes under study;
(d) the intra-household correlation between individuals for the variables under study;
(e) whether an alternative frame exists.
The population not living in households, as normally defined, includes the institutionalized and the homeless. This population may be of significance in a social survey of health, education, or general quality of life.

If the subset of the population to be covered is a large one, for example, adult women (fertility) or all persons over six years of age (literacy), then the household may provide a convenient sample unit, with total coverage of all eligible respondents in selected households providing unbiased estimates for the subset as a whole. There remain, however, difficulties with this approach. The precise number of respondents is not known in advance of the enumeration. Even if a listing of household members exists from an earlier visit, the household composition may have changed—one of the eligible members may be absent, temporarily or permanently, or another may have arrived. The possibility of unrecorded non-response exists and is difficult to check. It may also be difficult to interview two members of the household independently of each other (see Chapter 8)—but this can still cause difficulty when the respondent has been selected as an individual and not through the household. For surveys of social attitudes and behaviour this requirement is important.

If the subset of the population to be covered is a small minority, or if the phenomenon under study occurs only rarely in the population, and the definition of the subset or likelihood of occurrence is linked with an attitudinal or locational characteristic, the household is likely to be an inappropriate sample unit. A survey of taxpapers, for example, would involve only a minority of the adult population in many developing countries. We once received a request to include a measure of the incidence of leprosy in a household survey. Such a frame is clearly inappropriate for such a rare, geographically clustered, phenomenon and one which results in many of the sufferers being located outside a household structure. Selecting respondents for the first of the above examples requires the use of a frame based on centralized administrative records. And for the second, a random sample from any stratification of the general population is inappropriate—what is needed in such cases are tracer studies commencing with known cases.[1]

Surveys into the characteristics of individuals which use the household as a sample unit normally interview all eligible respondents in the household. The efficiency of this type of design for the main variables of interest will depend on their intraclass correlations as measured for individuals in the same household. Some demographic variables have very low intraclass correlations, but child nutritional status is likely to show significant correlations between individual children within a household. Any improvement of the design by subsampling individuals within a household and increasing the sample of households is, however, likely to be small. Once the household has been located, the individuals identified, and the survey objectives explained, the incremental time for interviewing all the eligible household members may be of minor significance.

A review of the above issues may lead the surveyor to consider a sample design that does not involve the use of the household: the question that then arises is whether there is an alternative frame that can be used to select sample respondents. It is the frequent negative response to this question that confirms the popularity of the household as the contact sample unit.

We summarize the position thus: the household provides a conveniently identifiable sample unit in many cases where the household itself is not the focal point of the inquiry, but there are instances where the use of a household unit just will not do, either because a significant proportion of the population of interest lies outside the household structure or because sampling households does not efficiently identify the individuals of interest. In these latter cases the absence of a suitable frame must be addressed head-on by arranging for one to be constructed—use of a household frame merely ignores the issue and does not resolve it.

12.2 SOCIO-DEMOGRAPHIC SURVEYS OF INDIVIDUALS

Individual, rather than household aggregate, data are required on such variables as: (a) fertility; (b) health; (c) education; (d) occupation; (e) migration. Most of these variables need to be assessed according to age and sex categories.

Population censuses provide the data for calculation of general birth and death rates. More detailed studies of fertility patterns, involving extended interviews, require a particular survey of women who have reached their reproductive years. A subset of these women are also the most common respondents for surveys of maternal and child health, including nutritional status. The available literature on these topics is extensive. More research has been undertaken into the most efficient way of interviewing women on fertility histories, and of interviewing mothers and measuring children in order to assess their health, than in most areas of data collection. The publications resulting from the World Fertility Survey programme are a rich source of guidance in all aspects of undertaking surveys in this area.[2]

Experience shows that properly conducted interviews do elicit the most detailed information on a topic many would be inclined to think would meet with respondent reticence and resistance. The particular issues facing a surveyor in designing a fertility survey include:

(a) definition of the age-range of women to be included—how young the lower age limit should be set, and whether women who have passed through their reproductive period are to be included;

(b) careful choice of the enumerator force (probably female) as maturity, sensitivity, and confidence are essential qualities;

(c) obtaining privacy for the interview—this may be difficult to achieve but failure to do so has a marked effect on the responsiveness of the respondents, particularly young ones;

(d) the desirability of linking a factual review of a woman's fertility history with an attitudinal and practice inquiry into family planning methods.

Careful decisions on the age-range of the population to be included are also required for surveys of educational levels. The questions will normally focus on the number of years of school attendance and the highest grade reached. If literacy is to be investigated, it is advisable to include a simple objective test, for example, a card containing a few simple sentences to be read aloud by the respondent. Without such a test the responses to questions regarding literacy generally result in exaggerated claims for the level of literacy of the respondent.

Occupation and employment status of individuals are sometimes viewed by the questionnaire designer as an uncomplicated, unambiguous topic. Such is far from the case as some of the examples of question sequences in Chapter 6 reveal. Superficial questions into these matters will tend to produce simple replies such as 'none' or 'farm work'. These answers will conceal a wide variety of real activities. Blacker poses the question, 'What constitutes work?' and claims with justification 'that anyone should regard the information as being easy to obtain would seem to indicate that that person is blissfully unaware of the problems involved'.[3] One problem concerns women who engage in a mixture of subsistence farming and domestic duties; another is the time reference period. Statements about activities within the most recent time span (whether a day, week, or month) have to be supplemented by a more general probe into the respondent's 'normal' occupation, although his work may be so casual that 'normal' ceases to have much meaning. We refer the reader back to the example quoted in Section 6.6, where we stated that by the end of the probe both enumerator and respondent would be completely befuddled.

Migration is a topic of growing importance in many countries as the pace of movement, particularly from rural to urban areas, quickens. A study of migration will not usually be suitable for a highly clustered sample because of high intraclass correlation. Problems associated with duration of residence and short-term residence in another location, discussed in the previous chapter in the context of the *de jure* population, recur in questions on migration. Popular questions on this topic, within the framework of a demographic survey, include the following:

(a) birthplace—country or administrative area, e.g. district;
(b) period in current residence;
(c) previous place of residence other than birthplace;
(d) place of residence one year previously.

Recall over long periods may be involved in these questions. The place (although often not the date) of birth is usually known and remembered. A more likely source of bias is political motivation to conceal or falsify one's place of origin. An out-of-date frame will result in a sample that under-represents recent arrivals in an area and increases the non-response resulting from departures from the area; frame up-dating is, therefore, very necessary in this context.

12.3 ATTITUDINAL SURVEYS

It is increasingly recognized that the stimulation and sustenance of development require not merely the provision of services and technology diffusion, but the structuring of these interventions to accord with the attitudes and perceptions of the population. Surveys aimed at establishing prevailing attitudes and perceptions are, therefore, becoming more common in developing countries.

One of the most tangible indicators of attitude to an available product or service is the rate of adoption by the targeted individuals. There is a close parallel between measuring this rate and performing commercial market research. Adopters can be subdivided into initial adopters (first-time buyers), repeat adopters, and consistent adopters (displaying brand loyalty). The negative equivalents are also of interest. For example, the proportion of those who had access to the product or service and its promotion but who declined to adopt it; and the drop-out rate—the proportion of those who tried the product and then rejected it.

Not only are these important indicators, but for the surveyor they offer particular advantages in terms of sampling and interviewing. As is well known the standard error of a rate derived from a sample is governed by the expression $\sqrt{pq/n}$ where p is the proportion adopting and $q = 1 - p$. Therefore, the required sample size for estimating p within moderate confidence levels is quite small—a fact that has permitted the extensive use of opinion polling in developed countries. Moreover, an interview and/or observation required to classify a respondent as an adopter or non-adopter is often both short and easy. But this is not always true as there may be problems in the definition of adoption. If the respondent accepted only part of a recommendation, or used a product at levels below what the suppliers consider necessary for it to be effective, is he an adopter? The answer depends on the use to be made of the resulting index. Even partial adoption may indicate a positive attitude, restricted only perhaps by economic considerations. On the other hand, if likely changes as a result of the adoption are to be assessed a more stringent definition of an adopter may be required. The use of adoption rates in different formats is of major importance in monitoring development projects and is discussed more fully in books on this subject.[4]

To probe attitudes and perceptions more fully requires a more detailed interview. We discussed some approaches to such an interview and their problems in Chapter 6. Attitudinal probes have been most fully documented in the context of fertility surveys and birth control practices.[5] They are of growing popularity in studies of the development process. The range of interviewing techniques available is wide. Participant observation studies, where a trained observer resides for a period in a community, participating in community life and using this opportunity to both observe and discuss issues with members of the community, have revealed deeper insights into individual and community reactions to development stimuli than emerge from one-off interviews by a stranger.[6] In the agriculture sector applied farm systems researchers have developed an approach combining an informal conversational phase by a small team who visit an area for days or weeks, with a follow-up, more structured interview, based on findings from the first phase, applied to a wider population.[7] Another school has emerged recently under the banner of Rapid Appraisal or Assessment that relies on informal contacts with individuals rather than structured interviews applied to a randomized sample.[8] Group interviews have been found useful in drawing out both community and individual reactions. All these techniques have their strengths and weaknesses and must be chosen according to the depth of insight that is sought. In general, the more structured interviews may reveal attitudes and perceptions at only a superficial level and run the risk that the respondent may give back to the interviewer what he thinks the interviewer wants to hear or what the interviewer appears to suggest. On the other hand, such a format can be applied by a large survey force over a widely dispersed sample. At the other end of the scale the intensive, unstructured participant observation study has the potential to reveal the real underlying attitudes of the people, but requires a skilled, experienced practitioner and can be conducted only on a case study basis.

12.4 DIETARY SURVEYS

Studies of food consumed by individuals are very specialized and are usually linked to a longitudinal medical research study. They should not be confused with nutritional status and household food consumption studies. Food eaten in the household is not equally distributed to each person at the meal, and attendance at meals may vary, household members sometimes being absent and visitors sometimes present. Whether in such a survey all individuals in the household can be included separately is another question. Successful dietary studies tend to concentrate on one type of individual— children, nursing mothers, etc.—although the household context can never be ignored. An individual's food intake over an extended period will be close to his energy requirement unless there is famine or continuous food shortage. Daily intake will, however, vary substantially from no food at all to an intake several times in excess of requirement. Accurate data are required, therefore, over a considerable period of time.

Whitehead, in the context of measurement of dietary intakes of individual children, summarizes four procedures as follows: (a) weighed food intake; (b) replicate diet analysis; (c) dietary recall; (d) maintenance of a diary. Whitehead's comments on these are summarized below. Weighing the food intake is theoretically the most accurate of the methods, but the disruption of the household may make the results untypical. Moreover, such intensive studies can usually last for only one week or less. The precise weighing method is particularly difficult to apply for foods such as porridge or stews. In such cases the creation of a replicate meal, matched both qualitatively and quantitatively with that eaten, gives the opportunity for detailed analysis. Plate waste must be allowed for. Twenty-four hours is the usual period for dietary recall and Whitehead comments: 'it is often thought that this method is only really suitable for use in sophisticated educated homes, but indeed the reverse seems to be true. The more complex the way of life, the greater tends to be the variety of foods consumed and thus there is more to recall.'[9]

Rutishausen reports no significant difference in results obtained using recall and dietary replicate methods.[10] She also used the diary method. Photographs of foods were shown as column headings on a piece of board with each day as a row. Pins were inserted by the mother to record the number of times each food was given to the child. She maintained this system on the same respondents for up to three years. The ability and dedication needed for such a study are clear and make a strong case for the case study approach.

12.5 LABOUR FORCE SURVEYS

The type of survey that investigates the employment, underemployment, and unemployment of the labour force is clearly important for developing countries with a rapidly expanding population and rural–urban migration. Despite the importance of these data, the methodology for carrying out surveys of this type is unfortunately still in the experimental stage.

The main reason for this is that the concepts and definitions of these varying degrees of employment are by no means clear. We have commented, in Section 12.2 above, that even an apparently simple question such as occupation may cause complications and ambiguities. The ILO has stated:

The concepts and definitions of employment, unemployment and underemployment applied which are based on the labour force approach developed in industrialized countries do not suffice for an adequate analysis of the employment problems.[11]

The ILO also singles out the choice of a suitable reference period as being a problem as yet unresolved, since 'The one-week reference period and the concepts used cannot sufficiently picture the real situation of economic activities.'

Few guidelines can be offered in view of these unresolved problems. First, the household is usually an appropriate sample unit for the reason stated at the beginning of the chapter. But a labour force survey may also fit into the category of a household survey as defined in Chapter 11, for the reason that the household is the unit of production as well as consumption. In this latter case the contribution of all household members, including children, must be included in the survey. But meaningful surveys into unemployment will need to encompass those who are temporarily outside the household structure such as recent migrants into towns who have no fixed abode.

Second, as Mueller states:

economists need employment data that have a longer time reference than the week or month on which conventional labour force surveys focus.... The year appears to be the most suitable reference period for employment studies.[12]

Third, productivity of the labour must be assessed. Mueller adds:

data on hours worked are more useful if they are associated with data on income from work, so that productivity per unit of time can be estimated. The point is that where work opportunities are insufficient work may be paced to fill the available time.

This brings us back to the issue of underemployment. We return to our oft-stated point that the majority of the population in developing countries are dependent to a greater or lesser extent on small-scale agriculture. To use a questionnaire stressing questions such as 'Were you seeking work last week?' or 'State number of hours spent on ... in last week' will not help much. A study of labour force and income for the agricultural population demands, in fact, a farm management type of study. Collinson and many others have conducted surveys of this type that reveal quite plainly the labour constraints according to season and the links between labour inputs and productive output. He writes:

Output is a measure of the performance of the system and quantifies the results of resource use at present levels of technology and management ability. It forms a benchmark for the evaluation of alternative improvements to be tested in the planning sequence. Enumerating the quantities of food produced also gives a basis for assessing the importance of each subsistence activity of the family.[13]

For that part of the urban population which is in formal employment in known businesses or services the conduct of a labour force survey is somewhat less difficult. Often, a sample is chosen from lists of workers maintained by selected firms. However, as mentioned above, most urban areas have a substantial itinerant casual labour force that is more difficult to capture. And a further important segment of the labour force is engaged in informal activities and individually run enterprises. A full survey, therefore, requires the careful establishment of a frame that goes beyond the wage rolls of registered businesses.

12.6 NUTRITIONAL STATUS SURVEYS

As we indicated in Chapter 1, one of the more promising developments in data collection in developing countries in recent years has been the conducting of nutritional status surveys of children based on anthropometric measurements of height (or length) and weight by age. These surveys have been promoted particularly by WHO, UNICEF, and FAO. Children under the age of six years are the group usually covered in such studies, because young children are the most vulnerable to nutritional deficiencies and because the indicators derived from the measurements are more specific and sensitive to change when related to this group. Some similar studies have been carrried out on adult populations, but the results are less definitive.

Three types of survey design are popular for accessing a sample of children. First, once again, the household is a convenient cluster sample unit even though the data are for selected individuals within the household. A nutritional status module is frequently added to a household survey commissioned for some other primary purpose. Not only is this convenient, but it does offer the possibility of presenting the results by certain household characteristics such as income or family size. The linkage of this module with a general household survey is made possible because experience in many countries has demonstrated that general purpose enumerators can be trained to take the required measurements, obviating the need for more specialized staff.

The second type of design is to use a community-level primary sample unit and record the anthropometric measurements for all children in the community. This is a suitable design for a survey that is intended to be repeated over time. In the Philippines and Indonesia children in selected areas are measured twice a year.

The third is to use clinics or schools as the points for capturing the sample of children. The staff of these institutions can be used to take the required measurements. The sample will not be representative of the entire child population as the children who do not visit a clinic or attend a school may well be from poorer families or remote areas. But as indicators for a large identified proportion of the child population they have validity, and movements in the indices over time may well indicate the direction of change in the whole population.

12.7   PRIVATE TRADER SURVEYS

We mention finally surveys of individual private traders. Strictly, data collection on trading activities falls within specialized surveys of businesses, but undertaking a survey of local market trading has certain features that fit within this chapter: no list of such traders is likely to be on record, their identification is likely to be missed in household surveys, and individual interviews, sensitively conducted, are required.

Small-scale individual trading in local markets is often engaged in on a part-time basis by many men and women whose primary occupations lie elsewhere. But there are many individual full-time wholesale traders who are potentially a mine of information on the marketing of local commodities if they can be accessed and their confidence obtained. In the absence of a frame, access can be obtained by sampling markets or trading centres and posting an interviewer at the market gate to identify traders as they arrive to sell or buy. This step is, of course, crucial, and the one when mishandled that will lead to difficulties. Traders will not fail to notice an 'official observer' making notes and suspicions will be quickly aroused. If they permit an approach to be made they are unlikely to reveal information that pertains to their operating margins even if they are fully licensed to trade, which is often not the case.

Our experience, however, leads us to be optimistic as long as the surveyor prepares the ground and is discreet. We have been surprised by how co-operative traders have been in surveys with which we have been associated even in circumstances where the movement of crops was technically forbidden.[14] Our message is simple: trader surveys, above all others, demand that the most patient preparatory work is undertaken and that the interviewers are able to establish confidence. First visits to markets must be to establish contact only, to introduce oneself and the survey. Frankness is of the essence—traders who cannot detect dissembling are a rare breed. Only when relations have been established—and patience is necessary—can the formal enquiries begin. This patient groundwork can be well rewarded. We have found traders to be amongst the most informative respondents we have interviewed, given trust on both sides. We have travelled with them on their arbitrage journeys and learnt more in a few days than more structured surveys on prices and volumes reveal over long periods.

Notes

1. Kalton, G., and Anderson, D. W., 'Sampling Rare Populations', *Journal of the Royal Statistical Society*, 1986, p. 149.
2. See World Fertility Survey Technical Bulletin series, International Statistical Institute: World Fertility Survey, London.
3. Blacker, J., 'A Critique of the International Definitions of Economic Activity and Employment Status and Their Applicability in Population Censuses in Africa and the Middle East', *Population Bulletin of the UNECWA*, June 1978, p. 47.
4. Casley, D., and Kumar, K., *Project Monitoring and Evaluation in Agriculture*, Johns Hopkins Univ. Press, Baltimore, forthcoming.
5. Stycos, J. M., *Putting Back the K and A in KAP: A Study of the Implications of Knowledge and Attitudes for Fertility in Costa Rica*, World Fertility Survey Scientific Reports No. 48, 1984.
6. Salmen, L., 'Listen to the People:Participant Observer Evaluation of Development Projects', World Bank, 1987.
7. Byerlee, D., Collinson, M., *et al.*, *Planning Technologies Appropriate to Farmers: Concepts and Procedures*, CIMMYT, Mexico, 1980.

8. Chambers, R., *Rural Development: Putting the Last First*, Longman, 1983.
9. Whitehead, R., 'Some Quantitative Considerations of Importance to the Improvement of the Nutritional Status of Rural Children', *Proc. Roy. Soc. Lond.* B 199, 1977, pp. 49–60.
10. Rutishauser, I., 'The effect of a traditional low-fat diet on energy and protein intake, serum albumin concentration and body weight in Uganda preschool children', *Br. J. Nutr.* 29 (2), 1973, pp. 261–8.
11. ILO, 'Recent Experience in Labour Force Sample Surveys in Developing Countries' in Brown, J., *et al.* (eds.), *Multi-purpose Household Surveys in Developing Countries*, p. 159.
12. Mueller, E., 'Design of Employment Surveys in Less Developed Countries' in Brown, J., *et al.* (eds.), *Multi-purpose Household Surveys in Developing Countries*, p. 189.
13. Collinson, M. P., *Farm Management in Peasant Agriculture*, Praeger, New York, 1972.
14. Casley, D. J., and Marchant, T. J., *Smallholder Marketing in Kenya*, Marketing Development Project (KEN 75/005) UNDP/FAO, 1975.

13

Data on Agricultural Holdings

'You have a few sheep in the paddock,' he said, 'Who attends to them?'

Silver Blaze

13.1 INTRODUCTION

In Chapter 11 we commented on the distinguishing feature of households in developing countries—they are units of production as well as consumption, small-scale farming being the way of life for most people. The collection of agricultural statistics, therefore, might be seen as an integral part of the collection of household statistics; indeed it is difficult to conceive of household surveys covering topics such as income, expenditure, labour, social conditions, and household enterprises being taken without reflecting the practices of, and returns from, farming. Unfortunately, an artificial distinction has often been drawn between socio-economic surveys and agricultural surveys. Some have questioned the inclusion of agriculture in a national household survey programme. Others have thought that the danger was the other way round; thus Zarkovic in 1962 stated: 'The impractibility is being recognized of efficiently collecting some types of agricultural statistics by using . . . household surveys.'[1] This is true for 'some' types; but we think it is undesirable to give primary importance to this particular differentiation in the development of a programme to investigate the economic and social conditions of the agricultural population.

The collection of agricultural statistics is an expensive undertaking because of the perceived necessity for frequent measurements and observations over an extended time period. Further, the methodology required to collect statistics of the production, marketed volume, and consumption of home-grown crops is by no means well defined. The consequence has been a comparative neglect of agricultural statistics at the national level, even when such statistics are of prime importance to the national economy.

Interest in statistics at the farm level has been more evident. Agricultural economists have been active in conducting farm management surveys in many countries with the stated objective 'to diagnose the problems confronting the farmer and devise more economic ways of using available resources'.[2] But, in the view of Collinson, the results of these surveys 'made little impression' during the 1950s and 1960s for various reasons including the 'ad hoc nature of data collection . . . and a failure to

see where and how widely any collected data could be validly applied'.[3]
Recently the collection of data similar to those included in the classical
farm management survey (see Chapter 14) has been undertaken as a
means of monitoring and evaluating agriculture and rural development
projects. This was an inappropriate choice and proved as disappointing
for this purpose as in earlier years. Individual researchers, too, in their
study of the development process need data on the agricultural sector and
have to struggle with the problems of definition and methodology.

As the machinery for planning grows more sophisticated the demand for
current data on the agricultural sector, at both the 'macro' and 'micro'
levels, also grows. In responding to the demand it is important that the
surveyor considers the options carefully, in order to find solutions to the
problems and to promote their adoption by co-ordinating his activities
with those of others working in this field. The following sections introduce
some of the issues.

13.2   THE HOLDING AND THE HOLDER

In everyday terms we talk of farms and farmers, with the farms divided
into fields. International definitions involve the concept of a holding and a
holder, with the holding divided into parcels and plots. These definitions,
not always easily understood at the local level, are based on the desire for a
standard unit, not subject to vague definition or widely different local
interpretations, that reflects operational control rather than legal ownership. The FAO defends the holding as a statistical unit thus: 'the
"holding" is generally different from the "establishment" normally used
as the statistical unit in censuses and surveys relating to industries other
than agriculture, although the two concepts may be equivalent in many
cases. . . . there is a pressing need to cover, to the extent possible,
agricultural production by all types of establishments regardless of the
economic activities under which such establishments are classified . . . the
holding is much more easily recognizable in terms of land and for
livestock, while the concept of an establishment is difficult to apply in
rural areas of most countries.'[4] A holding is defined as a unit of
agricultural production comprising all the land used completely or partly
for agricultural purposes and all livestock kept and operated under the
management of an individual or group without regard to legal ownership.

It is important to note that the land should be in use for agricultural
purposes or, having once been in use, be lying fallow with the intention of
being used again at a later date. The holding normally includes the land
occupied by the farmhouse, farmyard, and farm buildings. Communal
land, land never used for agricultural purposes, and natural forest are not
part of an agricultural holding. The practice of shifting cultivation, that is

the clearing and use of land previously under natural vegetation followed by its abandonment with no intention to return to it, is common in some parts of the world. The total area of a holding in such a case is the sum of the land areas in current use and excludes the abandoned land. Although communal land is not part of the holding, livestock owned by the holder but grazed on such land is; the distinction is simple enough but the practical difficulty lies in ensuring that such cattle are, indeed, included (see Section 13.6 below).

The holding is made up of one or more parcels, a parcel being a piece of land operated by the holder that is completely surrounded by land belonging to other holdings or to no holding at all. In other words, the parcels that make up a holding are separated from each other—the 'barrier' may be as little as a river or road, or the parcels may be separated by many miles of land not included in the holding. The parcel, in turn, is made up of plots, a plot being a piece of cultivated land containing a single crop or a single homogeneous mixture (see Section 13.6 below) of crops. Note that there is no necessity for plots to be divided by fences or even paths. The dividing line between two growing crops, even if unmarked in any way, is also the dividing line between two plots.

In any survey of agriculture it is important that these definitions, if used, are clear to the enumerators and are rigidly adhered to. There is a tendency, despite lengthy training, for enumerators to revert to preconceived concepts, so that a plot becomes synonymous with a field, that is, land demarcated in some way, and the holding is taken to include all land that the respondent owns whether or not he is operating it. Different interpretations of the definitions by various enumerators can make a nonsense of any subsequent classification of holdings by size, or by utilization. If the definitions as applied in a country are so 'alien' in concept that the enumerators are unable to place them in the context of local practice, then the local adaptation of the international definition has been badly done. In particular, in some regions, the area of the 'active' holding changes during the year, as well as from year to year. The definitions used must 'make sense' to the enumerator and the holder and this implies adapting the international definition to fit local customs rather than vice versa. Uniformity of interpretation within the nation must be sought, but is not easily achieved. Detailed knowledge of local practices, and of variations within the area surveyed, is required—another illustration of a topic where case studies can be valuable.

A sample of households is commonly used as the basis for an agricultural survey. It is necessary to identify the holder from amongst the household members and this stage is crucial for, using such a procedure, the holder is identified first and his holding later, during the interview. The household list acts as a substitute for a frame of holders or holdings because these rarely exist.

A household may contain no holders, one holder, or more than one. In most countries the great majority of rural households will contain one holder, so that there is a one-to-one relationship between household and holding, as reported by Panse following a global review.[5] Non-agricultural households, of course, can be removed from the agricultural survey without causing sampling problems. The more difficult problem arises when one household contains two or more holders.

The holder is the person who exercises control over the operations of the holding and is responsible for the utilization of available resources. He is not necessarily the owner of the land on which he conducts his agricultural operations—he may be a tenant, squatter, or occupier under customs of traditional usage. The corollary to this is that the legal owner of land may not be the holder in terms of responsibility for the agricultural operations. Holders should not be identified casually. Many apparent holders are not truly so. The holder must be taking the responsibility for the production of the crops and their disposal. A member of the household, other than the head, may appear to be operating a piece of land independently of the head, but when it comes to the disposal of the produce the underlying responsibility of the household head emerges. The holder may not take day-to-day decisions about the operation of part or all of his holding because he is satisfied with the way his wife or son is acting on his behalf. He would interfere if he became dissatisfied with the operations carried out. For example, if the wife of the head of the household omits to weed the maize on a piece of land for which she appears to be taking operational responsibility the head may instruct her to do so. In such a case it is the head of the household who is the holder. To quote Panse, 'It is the responsibility of decision-making that determines the operator.'[6] The delegation by the holder of authority for day-to-day decision making can extend outside his immediate family to a hired manager. The hired manager may, however, be considered to be exercising such authority in his own right so that he is considered to be the holder. This equating of the employee with the holder must be adopted with caution. In most cases the paid worker cannot be regarded as the holder—by definition his employer has the real control over the direction of his labours. Only if the manager is sharing in both the risks and profits of the enterprise can he be properly regarded as the holder, or rather, a joint holder.

If care is taken in the identification most rural households will be found to contain only one holder. Many households that appear, at first sight, to contain more than one holder will, on further examination, usually be placed in the simpler category. Indeed, we suggest that in some of the remaining cases, where the presence of two holders is clearly established, the household may have been defined incorrectly; what was listed as a single household is, in fact, two—with a holder in each. The FAO in describing and explaining the definition of a holder states

Agricultural operations carried out and commodities produced by different members of a household will normally be sufficiently pooled so that there is only one holder. . . . it will usually be necessary to determine one holder in each household in many of the developing countries.[7]

Nevertheless, genuine multiple-holder households do of course exist. In these cases, each holding will be enumerated separately, and care is required in putting questions that refer to the household—such as the history of its members or of sources of income other than farming. Either the household must be divided into sub-units attached to each holder, or the entire household (including the other holders) must be shown against one holder, with a zero entry on the household schedules of the other holders. In this way double counting may be avoided.

The opposite case occurs when two or more persons belonging to different households jointly operate a single holding. If one of these households is included in the survey the holding will be included but with an indication that it is jointly operated with others, so that appropriate account of the situation can be provided for in analysis. The holder can also be a corporate body, a co-operative, an association of growers, or a commune. No household details would be attached to such a holding. Members of these bodies may also be ordinary holders and the private holdings operated by them should be covered in the normal manner. The joint operation should be covered separately in terms of enumeration and tabulation. With these exceptions, there is no incompatibility between household and holding surveys: in fact, the household and holding are so frequently inseparably linked that surveys into one cannot afford to ignore the other.

Once the identity of the holder has been established the holding may be identified. Practical problems in this identification and enumeration include two issues which deserve separate treatment:
(a) how to enumerate land operated by the holder far from the place of interview; and
(b) what size limitations to impose.

If the agricultural survey is based on a sample of households, part at least of the holding may lie around the home, but the holder may also operate parcels of land in other places—perhaps in a different district or region of the country (even, but rarely, outside the country of his residence). These parcels are part of his holding, as defined above, but it may be impractical for the enumerator to visit them to make objective measurements of crop areas and counts of livestock. If this is so, it follows that the respondent also visits these parcels rarely, and is likely to have a wife or manager resident there. In many such cases it may be simpler to regard this person as the holder of the relevant parcel, thereby removing it from the respondent's holding and creating a separate holding. If this practice is adopted all enumerators must be trained to identify such 'caretaker' holders, in order that they have a chance of being included as a respondent. Instructions to this effect may seem to cut across those required to fix the definition of the holder in the way discussed above. Thus, confusion can occur with the attendant risk of double counting such a parcel or omitting it altogether. In countries, such as Ghana, where many holders live in one region growing a miscellany of food crops, but own a cocoa 'farm' in another region, the failure to adopt, and ensure the implementation of, a consistent approach by enumerators in both regions can invalidate the agricultural data collected. If the respondent, having no manager on his distant parcel, claims to operate it himself by making frequent visits, it should be possible for the enumerator to accompany the respondent to it, as it is likely to be within walking distance.

The common error is to identify the holding as being merely that land operated by the holder in the area of enumeration, or to fail to probe for the existence of parcels some distance from the place of enumeration. In either of these eventualities, serious underestimation of agricultural production may result.

Finally, is there a minimum, or indeed a maximum, size of holding for the purpose of the survey? A decision on this will depend on the precise objectives of the survey, but general guidelines can be offered. In practice a minimum size will be adopted, even if this minimum is the lowest size at which it is possible to measure and record the area. If enumerators are recording to one decimal place of a hectare, an agricultural operation of size 0.04 hectares or less will be too small to include. Normally, the surveyor will set the limit somewhat higher than this in order to exclude gardens of non-agricultural households that are used for decorative purposes and/or the cultivation of a very small amount of food for household consumption. If such a minimum is set it should be at a low level. Very large numbers of rural households engaged in subsistence agriculture operate holdings below one-half of a hectare. Because of their numbers, their elimination from a survey aimed at estimating total agricultural production may result in serious underestimation. In South Korea, livestock holdings were excluded from a sample census, and in Brazil, all subsistence holdings were excluded. Exclusions such as these may have to be made because of resource constraints, but they will naturally limit the usefulness of the resulting data.

An upper limit may also be appropriate for practical reasons. A questionnaire designed to be applicable to the majority of the agricultural population, namely smallholders, may be inappropriate for very large farms run on a commercial basis. Information on holdings at the upper end of the size distribution may be collected more efficiently in a different manner from that used in the main survey. Mailed questionnaires, or visits by senior survey staff with a specially designed questionnaire, are alternative approaches. If large holdings are excluded altogether, the survey report must make this very clear—especially as a repeated footnote to appropriate tables. Users may be grossly misled if the omissions and their importance are not well documented and thrust in front of them.

13.3 LAND MEASUREMENT AND CROP AREAS

Almost any survey of an agricultural population will be concerned with the extent and use of the land available to the holder and his household; so the areas of the holding and the area under each crop are likely to be required. Unfortunately, the systems of agriculture prevailing and the lack of numeracy on the part of the respondents may require measurement of areas during the survey. If the holder owns a title to the land he will probably know the area covered by the title, but he may not know the areas under cultivation. Moreover, the land covered by legal title may not be equal to his holding. Part of his land may be rented to others or used for purposes unconnected with agriculture. Or he may operate land not covered by the title but which is part of his holding. If the respondent cannot provide reasonable estimates of area, the surveyor has a range of techniques available including the enumerator's eye estimate, objective ground measurements (using one of various methods), and aerial photography. There is little need to dwell on the inaccuracy of enumerators' eye estimates. The enormous biases that these can introduce are well documented by Zarkovic[8] and others; although, owing to disappointment at the results of surveys using more objective methods, Zarkovic is one of many who have continued to look for ways of improving eye estimates. Certainly the very much greater cost of objective area estimation must result in a much higher level of accuracy if it is to be justified. But even the rare person with the ability to make reasonable eye estimates needs a reference point, that is, knowledge of the area at some point earlier in time.

Objective measurements of areas can be made in many ways. One frequently used technique involves triangulation and the fitting of a polygon to the shape of the area. If the angles of the sides of the area are to be measured a compass is used. Measurements of distances are made by pacing, or use of a measuring tape, a chain, a road measuring wheel, or a range-finder. The need is for a method that combines speed and accuracy, but unfortunately these objectives are conflicting. Triangulation by the eye of the enumerator, with lengths of the base and height of each triangle recorded in paces, is probably the quickest method, and requires only simple calculations. Although once commonly used, this method is not very accurate, and the recording of bearings and distances of the sides of the area is now recommended. The use of a measuring chain is probably the most accurate method of recording distances (within the present context, which excludes professional land surveyor techniques), but it is very time consuming. In our experience, the method that is both accurate and reasonably quick involves the use of a prismatic compass and a road measuring wheel. It is sometimes argued that the use of a wheel is impracticable in muddy or uneven terrain. We have found that, used by a well-trained enumerator, the compass and wheel method will give satisfactory results even in such conditions.

The advent of programmable pocket calculators has made possible the quick calculation of areas by enumerators and by the supervisors during field visits. The calculator can be programmed to display not only the area but also the size of the 'closing error'. Examples of programs and a discussion of methods of distributing the error around the polygon are given in an FAO Report.[9] A limit to the size of the error that can be tolerated without remeasurement is a matter of choice. The object is not to repeat the enumeration when the closing error is due to small random errors in individual measurements, but to detect single errors, large in magnitude, that distort the shape and area of the plot. Hunt suggests that the acceptance/rejection limit be placed within the range 5 to 10 per cent.[10] We recommend that the limit should be placed at the lower end of this range.

If pocket calculators are not available for the supervisors' use, the bearings and distances must be plotted onto squared paper by the enumerators and the area calculated by summing squares. If the plotting is delayed, the almost inevitable high error rate in taking compass bearings and distances will be detected only some time after the original measurements are taken. By the time the enumerator can be told to repeat the measurements he may be away from the area, or the crop area concerned may have been harvested. If the enumerator plots his measurements he will detect large errors and can then retake the measurements without delay. A further check in the office will then be mainly intended to correct slight errors in calculation, rather than to obtain measurements for bearing and distance.

Figure 13.1 shows an example of the type of form that can be used to record the bearings and distances.

If the budget and available facilities permit (or an aerial survey has been carried out for some other purpose), low-level aerial photographs can be used to obtain the areas of holdings and of plots that are of a reasonable size. It remains necessary for the enumerator to visit the holding in order to identify the plots belonging to the holding and to record the crop composition and other facts that may be needed. The enumerator can mark the plot, if identifiable on the photograph, the area of which can then be calculated in a properly equipped office. The enumerator should measure plots that are too small to mark on the photograph, plots that

FORM 12

AREA MEASUREMENTS

SHEET NO ....OF..............

ZONE	DIST	LOC.	SUB-LOC	E.A.	FARM	CARD	ORIG - 1 RECALL - 2
1 2	3 4	5 6	7 8	9 10	11 12	13 14	15

FARM NO ....

SUB-LOCATION ....

DISTRICT...............MONTH...............

LAND UNIT NO ....

LAND UNIT	CROP (1)	Y R.	R N S	CROP (2)	Y R.	R N S	CROP (3)	Y R.	R N S	A - B F.B.	A - B METRES	B - C F.B.	B - C METRES	C - D F.B.	C - D METRES	D - E F.B.	D - E METRES	E - F F.B.	E - F METRES	F - G F.B.	F - G METRES	G - H F.B.	G - H METRES	CNT.
16 17	18 19	20	21	22 23	24	25	26 27	28	29	30 31 32	33 34 35 36	37 38 39	40 41 42 43	44 45 46	47 48 49 50	51 52 53	54 55 56 57	58 59 60	61 62 63 64	65 66 67	68 69 70 71	72 73 74	75 76 77 78	79
										B.B.	DIFF.	B.B.	DIFF.	B. B.	DIFF.	B. B.	DIFF.	B. B.	DIFF.	B. B.	DIFF.	B.B.	DIFF.	X

LAND UNIT NO ....

LAND UNIT	CROP (1)	Y R.	R N S	CROP (2)	Y R.	R N S	CROP (3)	Y R.	R N S	A - B F.B.	A - B METRES	B - C F.B.	B - C METRES	C - D F.B.	C - D METRES	D - E F.B.	D - E METRES	E - F F.B.	E - F METRES	F - G F.B.	F - G METRES	G - H F.B.	G - H METRES	CNT
16 17	18 19	20	21	22 23	24	25	26 27	28	29	30 31 32	33 34 35 36	37 38 39	40 41 42 43	44 45 46	47 48 49 50	51 52 53	54 55 56 57	58 59 60	61 62 63 64	65 66 67	68 69 70 71	72 73 74	75 76 77 78	79
										B. B.	DIFF.	B.B.	DIFF.	B. B.	DIFF.	B. B.	DIFF.	B. B.	DIFF.	B. B.	DIFF.	B. B.	DIFF.	X

LAND UNIT NO ....

LAND UNIT	CROP (1)	Y R.	R N S	CROP (2)	Y R.	R N S	CROP (3)	Y R.	R N S	A - B F. B.	A - B METRES	B - C F. B.	B - C METRES	C - D F. B.	C - D METRES	D - E F. B.	D - E METRES	E - F F. B.	E - F METRES	F - G F. B.	F - G METRES	G - H F. B.	G - H METRES	CNT
16 17	18 19	20	21	22 23	24	25	26 27	28	29	30 31 32	33 34 35 36	37 38 39	40 41 42 43	44 45 46	47 48 49 50	51 52 53	54 55 56 57	58 59 60	61 62 63 64	65 66 67	68 69 70 71	72 73 74	75 76 77 78	79
										B. B.	DIFF.	B. B.	DIFF.	B. B.	DIFF.	B. B.	DIFF.	B. B.	DIFF.	B. B.	DIFF.	B. B.	DIFF.	X

LAND UNIT NO ....

LAND UNIT	CROP (1)	Y R.	R N S	CROP (2)	Y R.	R N S.	CROP (3)	Y R	R N S	A - B F. B.	A - B METRES	B - C F. B.	B - C METRES	C - D F. B.	C - D METRES	D - E F. B.	D - E METRES	E - F F. B.	E - F METRES	F - G F. B.	F - G METRES	G - H F. B.	G - H METRES	CNT
16 17	18 19	20	21	22 23	24	25	26 27	28	29	30 31 32	33 34 35 36	37 38 39	40 41 42 43	44 45 46	47 48 49 50	51 52 53	54 55 56 57	58 59 60	61 62 63 64	65 66 67	68 69 70 71	72 73 74	75 76 77 78	79
										B. B.	DIFF.	B. B.	DIFF.	B. B.	DIFF.	B. B.	DIFF.	B. B.	DIFF.	B. B.	DIFF.	B. B.	DIFF.	X

ZONE	DIST	LOC.	SUB-LOC.	E.A.	FARM	CARD	ORIG. - 1 RECALL - 2	HECTARES (ACRES ......)	NO. OF COFEE TREES	NO. OF TEA TREES	NO. OF CASHEW NUT TREES	NO. OF COCONUT TREES
1 2	3 4	5 6	7 8	9 10	11 12	13 14	15	16 17 18	19 20 21 22 23	24 25 26 27 28	29 30 31 32 33	34 35 36 37 38

FIG. 13.1 *Source*: as in Fig. 6.1

are concealed on the photograph by covering foliage, and land areas that are on sloping ground.

The methodology for measurement of crop areas is often designed on the assumption that the measurements can be taken on all holdings in the survey within some very limited period of time. In countries with definite summer and winter seasons this may be possible. However, in many countries the climate may be such that planting and harvesting is not only possible in two distinct seasons, but may overlap even within one season. Hunt drew attention to this type of difficulty when he wrote 'Practice in temperate countries is still all too often taken as the pattern.'[10] One of us has reported elsewhere as follows:

in one district of Uganda millet was planted from January until the end of May. Harvesting of the millet commenced in April and continued until October. In another district, groundnuts were planted from March until May and again from August to October. Harvesting commenced in July and continued until September, starting up again, after a pause, in November. Even on individual holdings it was noted that a period between the completion of planting and the commencement of harvesting did not exist.[11]

Continuous monitoring of land use, including the repeat use of one piece of land, is possible if regular visits are paid to the respondents throughout the agricultural year, suggesting a multi-visit survey involving interview and measurement (see Chapter 4). The implication of this approach is that the sample will be relatively small. A surveyor operating on a wide scale can also consider the possibility of visiting a much larger sample adopting a single visit method, and spreading the enumeration throughout the growing season. In this approach, crop areas are recorded as they exist at the time of the enumeration. The aggregate area under a crop can be estimated from data collected in this way, but tabulations of areas by characteristics of the holding are not possible.

The use of a single piece of land on more than one occasion during an agricultural year complicates statistics on total crop areas and on land under cultivation. The total area under crops will of course exceed the area of land under cultivation if the double use of a plot is reflected in the crop statistics but not in land utilization statistics. The calculation of land use statistics is further complicated by the practice of mixed cropping covered in the next section.

The surveyor's problem is clear. The collection of land use and yield data in complex farming systems involving very small plot sizes demands a highly clustered sample in order that one enumerator can maintain observations on a closely grouped sample of farmers: such a sample design is very inefficient for many agricultural variables that exhibit high intraclass correlations (see Chapter 4). We return to this dilemma in Section 13.5 on crop yield and production estimation.

13.4 MIXED AND ASSOCIATED CROPPING

To be measured a plot must first be defined. In Section 13.2 we stated that a plot contains a single crop or a single homogeneous mixture of crops. Two or more crops interplanted in the same land (mixed cropping), and a seasonal crop growing underneath a permanent tree crop (associated cropping), are common agricultural practices in the developing world. For example, in Ghana 84 per cent of the area under seasonal crops contained a mixture of crops. In Botswana 90 per cent of the area under millet and more than two-thirds of the area under sorghum contained other crops. Area statistics need to be treated with great caution, unless the method of area allocation for mixed crops is specified.

The difficulty is caused by the user's desire to have area statistics by crop: the area (and yield) of maize, rice, beans, and so on, each shown separately. But if most of the maize is grown mixed with crops such as beans or millet, this simple objective is difficult to realize. Various recommendations exist for allocating the whole or part of the area to the various constituent crops of a mixture. One method divides the area among the constituents without exceeding the total area of the plot. Another allows the entire area to be allocated to the main crop of the mixture with fractions of the area additionally allocated to the minor or subsidiary crops. No solution is completely satisfactory. The FAO reports:

Very few countries were able to follow the 1970 Programme recommendations to estimate the area which each crop would have covered if it had been grown alone; for example, the U.S.A. asked for actual area occupied by each of the associated crops. In Canada and India area was apportioned to each of the mixed crops. In the case of Liberia the area was apportioned to the first two important crops only. . . . In Turkey also the area was apportioned to each crop, if feasible, otherwise the total area was reported under 'others'. A large number of countries adopted an easy way out by allocating the entire area to each crop in the mixture.[12]

The Report continues to describe various methods used country by country. Most attempts at allocation of areas among constituent crops require a subjective eye estimation by the enumerator. The criteria on which he should do this are not clear. Is it the density of planting? Or the likely yield? Coconut palms may occupy very little of the area compared to the yams growing underneath them—but the palms may have been planted at the recommended spacing whereas the yams were planted scattered in a haphazard manner. Collinson concludes that 'any form of acreage equivalent is unsatisfactory. . . . The use of equivalents distorts land/labour relationships and prevents the comparative evaluation of productivity of what are effectively different enterprises'.[13]

A better solution lies in presenting area and yield data according to the reality of the agricultural systems in operation. The maize area, for example, should be shown as made up of the respective areas of maize grown in pure stand, and maize grown in the most common mixtures. The choices of which mixtures to classify separately can be decided from the result of a pilot survey or from the knowledge of agricultural agents. To speak simply of an area and yield of maize, when maize is normally grown interplanted with one, two, or even three other crops, is to mislead. The yield of maize may be little affected in some mixtures and seriously reduced in others. If the area and yields are recorded for each crop, classified by pure stand and various categories of mixtures, production can be calculated for each crop by aggregation. This is the statistic that is needed. Once again our plea is for simplicity. Record what is objectively seen. The basic tabulations should reflect this reality. Derived calculations at a later stage can remove the complexity, reducing the information to the simple digest required by the user who is not concerned with the agricultural systems in detail, only with aggregate estimates by crop. At the very least, the enumerator is spared the unreasonable task of applying vague criteria, to an imperfectly observed situation, in order to make an almost impossible decision.

There are subcategories of mixed cropping that the surveyor needs to be aware of, lest his area estimates turn out to be meaningless. Examples are relay cropping where a sequence of crops are planted in the same field, but at different times within the growing season, and continuous cropping where a crop is harvested on a piecemeal basis to provide just a sufficient supply for home consumption, with replanting of the harvested sections of the plot taking place almost simultaneously. Relay cropping commonly occurs with mixtures of cereals. Continuous cropping is mainly observed with root crops such as potatoes and cassava.

13.5 CROP YIELDS

The estimation of crop yields presents as many problems as the measurement of areas. First, is it the intention to measure the biological yield or the economic yield? If so, then direct harvesting methods must be used. The most popular involves the demarcation of small areas within the plot, and the harvesting of these small areas by the enumerator who retains the crop(s) harvested for drying and shelling and then records the yield. What size of 'crop cut' should be taken? How is bias in harvesting the sample 'cut' to be avoided? How is border bias in choosing the sample areas to be prevented? (Selected random co-ordinates may result in an area that cuts the edge of the plot; depending on the method used to avoid this problem, a bias may be introduced owing to under-representation of edges of the plots.) The treatment of mixed cropping has also to be determined. If one crop of the mixture is ready for harvest, but not the others, is it possible to harvest just the one that is ready?

If the economic yield, that is, holder-harvested production, of each crop is to be recorded, many of these problems disappear, to be replaced by others. What frequency of visit is required to 'inspect' the harvested crop? Is the holder able to remember and report the harvest in quantity terms? If so, what recall period is appropriate? How is the holder's response in non-standard units (baskets, etc.) to be converted into standard units of quantities? How is the amount harvested piecemeal for casual consumption prior to the main harvest to be estimated?

These issues cannot be discussed in depth here. The serious student should refer to the extensive literature on this subject, including the works of Mahalanobis, Sukhatme, Zarkovic, and others—Zarkovic being an excellent starting point. A few guidelines are offered stressing the practical issues. There is no dispute that crop cutting is an extremely laborious procedure that nevertheless often results in grossly erroneous estimates. Evidence is accumulating from recent trials that the bias is usually in the direction of overestimation by as much as 14–20 per cent of the measured total harvest.[14] This objective method has much to recommend it provided the enumerators are well trained and highly motivated, and that it is conducted with great care taken regarding the use of random co-ordinates, with a valid procedure for dealing with selected areas that fall partly outside the boundaries of the plot, and with precision harvesting of the selected areas following rigid rules about the inclusion of rows near the border. Usually these conditions are not fulfilled. However well trained, the enumerator has a subconscious reluctance to accept the dictates of random co-ordinates if the resulting area is clearly 'untypical'. For example, only rarely will an area be recorded with no plants inside it, and hence a zero yield. A cursory examination of many haphazardly sown plots on irregular ground will show that there are such barren areas that should be reflected in zero value crop cuts. Human nature also simplifies complicated methodologies for dealing with the border problem by pretending that the random co-ordinate at issue lay slightly further inside the plot. The tendency to include in the crop cut plants on the edge of the demarcated area, that are in reality fractionally outside, is well documented.

Estimates of yield from objective measurement of sub-plot crop cuts may be best left to agriculture research stations and specialized surveys that specifically require estimates of biological yield. Even in these applications some recent results are disturbing.[15] For general agriculture survey work we do not recommend it, based on the following argument. Crop yields exhibit high intraclass correlations; objective sub-plot harvesting requires a static enumerator and a highly clustered sample which is, therefore, a very inefficient design; if this price is to be paid it must be justified on the grounds that the non-sampling errors are almost eliminated; the evidence is that in most cases this is not so.

We also question whether in most non-specialized surveys it is *biological yield* estimates that are required rather than holder-specific *economic production* estimates. The estimation of the holder's actual harvest is in one sense much simpler: what is the number of bags, baskets, or whatever that the holder brings to his barn or store? In practice, serious difficulties arise. Usually the crop is harvested in a form that is not that required for the yield estimate. For example, maize yields are required in terms of grain, not in terms of number of cobs, which is all that can be reported at harvest time. The harvesting period may extend over several days or even weeks, and the enumerator cannot then be present throughout; by the time he returns to the holding, part of the harvest may already have been consumed or sold.

Certain crops are easier to handle than others. Crops such as coffee, if it is laid out to dry in the holder's compound, can be measured by the enumerator, even if the visit is some time after the harvest. Yams, large and individual, can be counted by the enumerator in the store. Indeed, such individual crops can be counted in terms of plants in the plot and an estimate of the number harvested obtained by visiting the plot and counting the remainder (although this assumes no loss or waste). The average weight per plant can be calculated from a sample in the store. Access to the store requires the co-operation of the holder—but without this all efforts at economic production estimation, by whatever method, are doomed to failure. Some surveyors report difficulty over the store, but our experience is that gaining access is less troublesome than estimating the quantity of stored crop inside.

Cereals give trouble, as they are often harvested 'green' and in a condition other than that needed for yield reporting. Samples can be taken for converting the crop as harvested into grain equivalent, but in many cases the enumerator actually relies more on the holder's estimates than the surveyor would wish.

How accurately can smallholders estimate their production? This is a contentious issue, but we believe that there is an unjustified tendency to assume that the farmer can tell us little of value in quantitative terms. Our own experience in several countries in Africa leads us to place more confidence in the farmer's estimate. There can be few crops more difficult to judge whilst still standing than mixed sorghum and millet, yet farmers in Nigeria in a controlled trial estimated total production as closely as the sub-plot crop cut method carried out by experienced enumerators.[16] Crop forecasts in Kenya in the late 1970s were based on farmer expectations and were found to compare favourably with later estimates based on the survey enumerators' returns. Surveys of cereals in the Philippines and Thailand have provided evidence that farmers' estimates are of reasonable validity.

The care taken in eliciting the farmer's estimate within an appropriate framework is all-important. Yields per hectare may mean nothing to most smallholders, but production from his holding, in a unit such as a standard bag, may be a different matter. Inevitably, the estimates will not be standardized for moisture content and so forth, but many of the so-called objective measurement surveys do not standardize the recorded weights either.

The potential gain from using farmers' estimates in terms of a less clustered sample design, and, possibly, a larger sample, is very great. The issue, therefore, can be clearly stated: will the improved accuracy to be achieved from objective measurements more than compensate for the price in sampling efficiency? We suggest that the evidence is that the answer is no. If the reader remains incredulous it at least behoves him to conduct a pilot test before adopting the objective measurement approach. Such a pilot involves testing the two methods independently conducted on a sample of fields which are then totally harvested so that a standard for comparison is available.

We have indicated earlier that reporting the areas of small plots is likely to be outside the farmer's ability. If production is to be estimated using interviews only, land utilization and aggregate cropping areas must be estimated in some other way. Sampling strips across the domain of study either by ground or air transects offer such an alternative. Strip sampling using low-level aerial photography has been found to be competitive in certain instances with the costs of objective field measurements. The use of ground transects[17] (pacing a fixed distance from a random starting-point along a selected compass bearing recording the number of paces for each land use) has become unfashionable. There are problems of timing, but these are equally present in the context of measurement of fields. Some old-fashioned techniques may still be of value in certain circumstances.

13.6 LIVESTOCK

An accurate livestock count is sometimes considered the most difficult agricultural statistic to obtain. Although the difficulties are numerous, we incline to the view that obtaining accurate crop information is more difficult. Three types of husbandry should be distinguished when the problem of counting livestock is being considered. They are:
(a) livestock normally grazed on, or near, the holding;
(b) livestock operated by the holder, but grazed on communal land, returning to the holding infrequently;
(c) livestock of nomadic peoples that are attached to no land holding.
Difficulty in obtaining good data on livestock is due either to the holder's not knowing the number he owns, or his reluctance to give away this

information. Both of these problems exist in certain areas and among certain respondents. There are holders who still do not assess their herds in terms of precise numbers. Much more commonly, genuine ignorance regarding the true number of small livestock, such as goats or poultry, will be found even among numerate respondents.

Reluctance to reveal numbers when known will arise partly because the holder regards these as private matters, and partly because he fears increased taxation. These problems are real, but not universal.

The solution to the first husbandry type seems simple enough: the livestock must be counted by the enumerator. If the cattle are on the holding, they are available for inspection (but there may be at least passive resistance to counting). Even if the holder knows the total number he may not be able to provide the numbers in the age and sex categories required by the survey. A visual review of the herd by the enumerator and respondent will provide this fuller information.

Difficulty may be experienced by enumerators in classifying animals by grade or breed. The use of terms such as 'improved' and 'unimproved', or 'grade' and 'local', must be defined, so that the resulting dichotomy is not only of value to the users, but can be applied by the enumerator. Strict veterinary definitions regarding the classification of second- and third-generation stock from cross-breeding pedigree and non-pedigree stock may set the enumerator too difficult a task. Our view, similar to that expounded in Chapter 11 with reference to the head of household, is that the enumerator may have to accept the holder's opinion as to the grade or status of the animal. The possible need to distinguish between 'working' and 'non-working' stock should not be overlooked. Oxen, camels, horses, donkeys, etc., can be vital inputs to the agricultural and marketing activity, and need to be classified as such.

The second category presents more difficulty, as the timing of the interview may not coincide with the infrequent presence of the livestock on the holding. If the survey is a single-visit survey, there is no alternative to accepting the holder's estimate, unless the communal grazing is nearby, in which case a search for the livestock can be made. But the holder may not reveal the existence of absent livestock and there may be no evidence that the enumerator can use to check that information is being concealed. Our experience, however, has been that, if the interview is conducted properly, with the correct conversational but probing approach, the existence of the livestock will usually be revealed. The incorrect recording of 'no livestock' is often at least as much the fault of the enumerator as it is of the respondent.

If the surveyor suspects that there is large-scale understatement of livestock ownership in these circumstances, close collaboration with livestock and veterinary officers may provide independent evidence. Officials dealing with livestock are more likely to *underestimate* than overestimate the extent of livestock ownership and of numbers owned, since they are more informed about cattle subject to dipping or vaccination campaigns. If the figures reported for a district or subdistrict are at or below the livestock officer's estimate, the surveyor should realize that his estimates may well be too low. A search should be made for detailed figures in local offices and time series developed where possible.

The livestock of nomadic peoples presents a different scale of difficulty altogether. Yet, in one sense, there is no difficulty at all, since the problem may be stated as one of finding the respondents rather than their livestock. If the respondents are located, their livestock will be found with them. Clearly, a survey of nomadic people—any survey—requires a tailor-made sample design, survey methodology, and enumeration technique. To attempt to include the nomadic sector of the population in a normal household survey is to invite failure.

Special concepts of household and holding are needed and the area cluster type of sample design is inappropriate. Indeed, as no frame may exist on which a sample can be based, and as a mobile population of people can scarcely be sampled in the same way as fish or birds, it may be questioned whether a sample survey is viable in these conditions. A census, and a seasonal itinerary of the pattern of movement, is a prerequisite for successful surveys of such populations.

If the pattern of movement is universal and regular in timing, for example, from north to south during the dry season, it is sometimes possible to enumerate the livestock by utilizing teams of enumerators moving in the opposite direction or even enumerators stationed at a (previously established frame) of water-holes at an intermediate point in the migration. Such an effort certainly requires the involvement of local officials, both technical and administrative.

Pending the availability of a frame, the study of nomadic populations (and their livestock) may be suitable for individual researchers collecting data at the case study level. If estimates of livestock numbers are needed for nomadic areas, independently of the holder and his household, low-level aerial photography may prove useful: it has been used (with varying success) for cattle and wild game counts.

The basic principle is to fly low-level, constant altitude, 'strips' across the area of study—the strips having been selected according to normal sampling theory. An observer makes a visual count within a defined width, supplemented by photographs taken at intervals. Western describes how this works:

In flight the counters will look out from the eye marker on the viewing window; through the wing rods which demarcate the strip width, into the ground. All animals which fall within the markers are counted, those outside are ignored. ... With herds over 15 to 20 animals it is usually necessary to take a photograph for subsequent accurate counting.[18]

A count made in this way will need to be supplemented by ground teams that check a sample of herds in order to verify the accuracy of the count of individuals within a herd and to classify them by type, age, and sex. Such a system is attractive in concept and labour saving in practice (ground counts have been estimated to take as much as fifty times more man-hours),[19] although involving high-cost man-hours and equipment. There are, however, many sources of bias. Western enumerates these as 'flying times and duration, aircraft bank, flying height, speed and strip width, counters' search-image and actual counting ability from direct and photogragraphic observations'.

The ability of the pilot to fly on a precise bearing and at a constant altitude and speed even over undulating ground is crucial: the flying height is usually two hundred or three hundred feet. Aircraft bank will affect the strip width. Early morning and early evening are recommended as the best times for counting animals[20]—factors, such as the shadow cast by the animal, affect their visibility. The largest potential bias arises from the observer's ability to count in the circumstances involved. One factor that will reduce his or the photograph's efficiency is the presence of considerable cover, in the form of vegetation, under which animals may be concealed. One thing may be said with confidence: if aerial counts are to be used the team involved must be very experienced—with the right hands and eyes useful results can be obtained in circumstances where large-scale ground counts are impossible.

Poultry counts will be approximate. The number of free-range poultry at any one time is probably unknown to the holder and cannot be ascertained with great accuracy.

Milk may be a basic food and an important income earner for the holder. Surveys to estimate milk production and disposal are becoming more common. The problems presented are similar to those encountered in estimating crop production and disposal. Total production rather than yield per animal is likely to be the level at which estimation is possible; even this may be difficult in a large survey. Milk yields, like crop cutting, are best estimated only in specialized studies.

Statistics of disposal of livestock and, in particular, births and deaths are also best collected in small-scale inquiries, and should help in the interpretation of general livestock statistics. When the estimation of accurate human birth and death rates has exercised the minds of demographers for many years, with no slackening of the debate yet in evidence, it is scarcely to be expected that accurate livestock birth and death rates will be obtained in a large-scale, single-visit survey.

Notes

1. Zarkovic, S. S., 'Agricultural Statistics and Multisubject Household Surveys', 1962, reprinted in FAO, *Studies in Agricultural Economics and Statistics, 1952–77*, Rome, 1978.
2. de Wilde, J. C., *Experiences with Agricultural Development in Tropical Africa*, Johns Hopkins Univ. Press, Baltimore, 1967.
3. Collinson, M. P., *Farm Management in Peasant Agriculture*, Praeger, New York, 1972.
4. FAO, *Programme for the 1980 World Census of Agriculture*.
5. Panse, *Some Problems of Agricultural Census Taking*.
6. Ibid., p. 22.
7. FAO, op. cit.
8. Zarkovic, S. S., *The Quality of Statistical Data*.
9. FAO, *Report on the 1970 World Census of Agriculture*, Rome, 1977.
10. Hunt, K. E., *Agricultural Statistics for Developing Countries*, FAO, Rome, 1977.
11. Casley, D. J., 'Problems in Estimation of Crop Areas and Crop Yields', *Bull. Intl. Stat. Inst.* 46, 1975, p. 533.
12. FAO, op. cit.
13. Collinson, op. cit.
14. Poate, C. D., and Casley, D. J., *Estimating Crop Production in Development Projects: Methods and Their Limitations*, World Bank, 1985.
15. Internal Communication from M. Greeley to Author referring to a crop loss measurement research study conducted in Bangladesh, 1986.
16. Poate and Casley, op. cit.
17. Ibid.
18. Western, D., *An Aerial Method of Monotiring Large Mammals and their Environment*, Project (KEN/11/526) Working Document No. 9, FAO, 1976.
19. Caughley, G., *Aerial Survey Techniques Appropriate to Estimating Cropping Quotas*, Project (KEN/SF/26) Working Document No. 2, FAO, 1972.
20. Norton-Griffith, M., *Counting Animals*, African Wildlife Leadership Foundation, Nairobi, 1975.

14

Monitoring, Evaluation, Surveillance, and Forecasting

The past and the present are within the field of my inquiry, but
what a man may do in the future is a hard question to answer.

The Hound of the Baskervilles

14.1 DESCRIPTION OF COVERAGE

Suitable data for four activities of growing importance are described in this chapter. The coverage of the fourth topic is clear, and stands apart from that of the first three, which are closely interrelated and need to be distinguished.

Monitoring and evaluation in developing countries are primarily concentrated with a single development project or a complex of interconnected projects. Arrangements for these surveys are now frequently requested by donors or project financiers and are sometimes included only at their insistence. The two activities shade into each other. It is difficult for anyone to evaluate unless information has been collected by monitoring the project as it proceeds; meanwhile information collected during the monitoring may have led the project management to evaluate its actions.

Monitoring may be defined as a continuous assessment of the functioning of the project in the context of design expectations. Are staff and materials available as required and are they being disposed effectively? Is the building of the feeder road proceeding according to timetable? Are the farmers using the fertilizer distributed? Are they putting it on the crops for which it was intended? Put in this way, the activity is an essential part of good management practice[1]. If resources are being provided, then any efficient management should make checks to see that they are:

(a) being delivered efficiently to the targeted population;
(b) being used as intended by the project design;
(c) having the effect they were planned to have.

Monitoring detects failures under these headings, so that management can take remedial action by adjusting and adapting procedures and allocations. It is often said that ensuring the provision of timely information is monitoring, and that the management feedback adaptation is the result of their evaluation of that information stream. However, when a project is proceeding satisfactorily within its own framework, and the management response is primarily to drive the project ahead and to make adjustments to ease frictions and equalize workloads, etc., it is simpler to regard the entire process as monitoring; and to reserve the term evaluation for two larger scale purposes.

The first relates to current or ongoing evaluation that takes place whilst a project is in progress. It arises when monitoring suggests that a major change in the project design is required or when replications of the project are under consideration. These involve larger issues than a managerial reallocation of resources within the existing project design. The second major type of evaluation occurs either at the end, or sometimes after the end, of a project's activities. Such an evaluation tries to answer much broader questions relating to the project's overall impact.

Evaluation then may be defined as a periodic assessment of the relevance, performance, efficiency, and impact of the project. This function will draw on the data available within the project that were needed by management to monitor progress, but will need to draw also on supplementary data that enable the performance within the project to be compared with that outside the project in terms of time, area, or population.

Another argument for drawing the line in this way is an administrative one. The extent to which monitoring and evaluation can and should be carried out by persons or institutions outside the project management has been difficult to resolve. Under our suggested distinction, monitoring would be carried out by the project management. In one sense this does little more than help to ensure that the project management is in fact following good management practice. If it is possible for someone or some organization to make a better job of it than the existing management, then they ought to be brought into the management. Once the project has been started all efforts should be directed to its success; available management talent should be inside helping it succeed, not duplicating the project's information arrangements and carping from outside.

Indeed, much of the information required for monitoring can only be derived by those working on the project. Outsiders trying to tune into internal lines of communication will not be able to do a good job, and will usually be resented. We have already stressed that administrative statistics are only likely to be reliable if the administrators require that information for their own purposes: similarly, monitoring data generated within the project will only be properly collected if the project management is using the data to help it do its own job.

Another aspect also falls into place if our distinction between monitoring and evaluation is adopted. Monitoring takes place within the framework of the project: it is done by the project management, and any results or discussion of it will form part of management's normal reports to those ultimately responsible. Evaluation will always require the involvement of persons outside the project management. Ongoing evaluation will normally be carried out by a joint insider–outsider group: ex-post evaluation may be a joint activity or it may be carried out entirely by persons not involved in the project management.

Evaluation involves comparisons outside the project for two reasons: first, in order to establish what part of the changes occurring in the project area can be confidently attributed to the project. This assessment requires additional information to that derived from monitoring. It will, for example, usually require surveys to establish the situation in the project impact area before and after the project. The changes occurring will have to be compared with changes in other areas in order to assess the 'real' effects of the project inputs (see the next section). Secondly, evaluation should assess how far other uses of the resources might have generated better results. The attributed effects will have to be compared with returns on other projects. Such a comparison will involve cost/benefit techniques, discussion of which lies outside this book.

The assembly of relevant project and nation-wide data and their successful analysis are tasks that a developing country, expecting to finance a number of development projects with bilateral or international assistance, could allot to a specially set up evaluation unit. Its appropriate place in the administration depends on individual circumstances; but it must be able to co-operate with donor agencies and local resources in universities, research institutions, and consultant agencies, and it will certainly need to work closely with the national statistical organization.

Surveillance is a general form of monitoring, relating not to a specific project or projects but to day-to-day aspects of a country's social and economic life. In a very general way, the continuous stream of information flowing up through local and central government units to the top executive and decision-making bodies monitors the general economic and social progress of the country. For example, the annual national account calculations monitor economic performance. More specific surveillance programmes may be introduced which watch over and serve as 'early warning' systems in areas where the ordinary channels of communication require supplementation. The topic that has received most attention in this way in several countries is nutrition.

14.2   CAUSE AND EFFECT

The literature of cause and effect is very extensive; and the following brief comments are introduced just to provide some background to the discussion of evaluation. In general, within a particular theoretical or expectational framework, a case that action X has had the result Y will be supported if:
(a) X occurs before Y,
(b) X has been followed by Y,
(c) possible causes of Y, other than action X, have been ruled out.

In the context of project design, X is the project and Y is the improvement (or package of improvements) the project is expected to achieve. The project will have been put in hand because the general theoretical or expectational framework suggests that doing X will result in Y. There will be no problem with (a) since Y will be assessed after X has been completed. What are required are data that throw light on (b) and (c).

So far as (b) is concerned, whether X has been followed by Y can be ascertained from a study of the information collected for monitoring, augmented by a comparison from appropriately designed surveys carried out in the impact area before and after the project. The actual design and implementation of such surveys are, of course, not as easy as that simple statement sounds. The effective ruling out of other causes—(c) above—is much more difficult, both in principle and in practice. Any improvement found may, for example, be due to general causes linked to the passage of time—the diffusion of technology, the rise in general education levels, a continued favourable climatic period. Since the impact area has been selected for the project, it may also be receiving relatively favourable treatment from a range of government and/or other institutions, and it may be that this overall assistance is the main cause of any improvement. It may even be that the improvement is due to the inhabitants in the impact area reacting enthusiastically to X, not because of what X is in itself, but because X is taken as a sign that at last government has shown special concern with the area's problems: that is, almost any sympathetic intervention could have sparked off the local actions leading to the improvement. Ways in which some of these problems might be eliminated can be usefully discussed in the framework of experimental design.

A standard design is represented by the accompanying diagram:

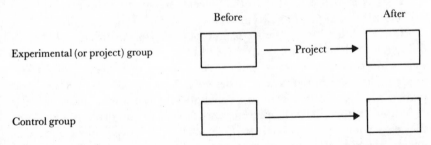

There are two groups, the experimental (or project) group and a control group that consists of comparable people. The 'before' situation of both groups is observed; the project is then implemented (for the project group only), and the 'after' situations assessed. The changes for both groups from before to after are then estimated, and the difference in the changes is attributed to the project. Suppose, for example, the project is to intensify farmer training. If the comparison of before and after situations shows an improvement of 30 per cent in yield in the project group, but an improvement of only 10 per cent in the control group, then the gap between the 30 per cent and the 10 per cent improvements might be attributed to the project, provided the monitoring information supported the argument.

We say 'might be', because the prerequisites for this experimental model cannot be met in reality. Human beings cannot be allocated at random to experimental and control groups. No control group can be found that completely matches the experimental group. Different degrees of matching can be attempted. A sample of farmers can be selected in the project area and characteristics of them, their households, and their holdings recorded: for example, one of them may be aged 45, farm 1.5 hectares, and have a household of 6. An attempt can then be made to find individual matches for the sample farmers in neighbouring areas, and both samples can be surveyed before and after. It is clear that matching of this detail is very difficult to set up and maintain. Further, if a sample of farmers are identified for this purpose before the project starts, they may well receive special attention from the project operators and changes in their before and after production will not then represent the true improvement in the project area.

Cruder matching procedures may be tried. Villages near the project area with, for example, similar frequency distributions of holdings by size may be used as a control group. The cruder the matching procedure, however, the less sure is the comparability between the experimental and control groups; and therefore the weaker the argument for the attribution of the effect of the project. If the experimental and control groups are not really comparable, then any divergence in the before/after changes in the two groups may not be due to the project, but merely the result of pre-existing differences in the two groups.

Another difficulty is that it is not possible to ensure that the only difference in the conditions experienced by the two groups is that one has the project and the other does not. During the period of project implementation—often years—other variations will almost certainly take place affecting the two groups in divergent ways; and it may be these variations, rather than the presence or absence of the project, that cause the dispersion of improved performances. The experimental model[2] nevertheless remains a useful conceptual scheme against which to consider the practical possibilities, to which we now turn.

14.3 MONITORING AND EVALUATION

We start with an emphasis on the overriding need for speed in data collection and analysis for these purposes. Two aspects are particularly

important. First, the data to be collected must be reduced to the minimum. We have cited this before as a general rule, and stress it now because evaluation surveys provide ideal grounds for proliferation of questions caused by two dangerous attitudes:
(a) the 'since we're there let's ask this' attitude (see Section 2.1);
(b) the 'play-safe' syndrome—the attempt to guard against neglect of a causal feature by covering any remotely related topic (see Section 2.1).
It is very difficult not to succumb to these temptations, but unless a determined effort is made to limit the extent of the data collected, masses of information will accumulate, most of which will never be analysed or used. The user–surveyor dialogue will have to put particular emphasis on this point; and it will have to engage the project management and the evaluation team (although the evaluation team and the survey group may be the same persons in some cases).

The second requirement is complete assurance about data processing capability. This is not restricted to formal computer availability. Some of the monitoring information will be in the form of progress reports and financial statements, many of which should be designed for rapid desk summary and analysis. Management must have timely information if it is to bring the project to a successful conclusion. Evaluation must also proceed rapidly, since decisions will normally have to be taken about further projects in the area or the desirability of repeating the project elsewhere. Since the full effects of a project (especially unforeseen unacceptable results) often take considerable time to make themselves felt, it may be necessary to make a first evaluation and provide for a second review after a further period. This need will be absorbed in a continuing programme of monitoring and evaluation if the project moves into a second stage, or a related project is mounted in the area.

Monitoring information for management will consist of returns relating to the arrival, disposition, and employment of resources (financial, equipment and materials, and staff), and current indicators of immediate results. The form and frequency of these returns and methods of collection and analysis should be designed primarily for and with the management. Any additional requirements suggested by the evaluation team should be closely scrutinized and, as a general rule, accepted only if the management team is willing to accept them, or can be convinced that they are worth the effort required.

The general framework within which this data stream is organized must also fit in with the surveys for evaluation. The following issues require decisions so that appropriate surveys can be mounted:
(a) definition of the 'impact' area;
(b) definition of the measures of impact;
(c) methods to assess the confounding effect of factors exogenous to the project;

(d) the time-scale required to measure trends rather than short-term effects.

The definition of the impact area, or the population affected by the project under evaluation, may be simple. The project itself may have selected the participants who receive the training, the services, or whatever is offered. In other cases, such as the provision of an access road or a health centre, the population to be surveyed is less easy to determine. It may be, of course, that the identification of who is affected and who is not is one of the aims of the 'after' survey. The impact area can sometimes be determined by an on-the-spot examination of the local topography, taking into account natural barriers, such as a river or forest, and other features of a man-made and administrative nature. Sometimes it may be necessary to conduct a pilot survey to establish the limits of the project area. In finally deciding upon the boundaries, the normal rules for selection of primary units should be kept in mind, particularly the need for the boundaries to be clear and unambiguous. Notional lines drawn on a map will always lead to confusion in the field.

The definition of the measures of impact may seem simple enough, given that the project has certain objectives. Certainly, when the project is mainly concerned with the diffusion of a particular technological innovation, some measures immediately suggest themselves. For example, if the project is designed to improve crop yields, the variable to be evaluated may be said to be defined. The provision of hybrid seed, fertilizer, or other inputs does not, however, automatically imply that they will be used for the purpose for which they were intended, since experience shows that farmers often fail to use these inputs, or use them on crops other than those for which they were provided. Monitoring should include study of the adoption rate and use of the inputs, as well as an initial assessment of production changes. The variables that are critical for evaluation will often go beyond the measured yields—although this will be the first priority. A provisional list includes economic indicators, such as income, expenditure, access to and use of markets, the disposal of home production, and consumption; social indicators, such as access to and use of essential services and utilities; health; education; and employment. It is the selection from within this wide list that is important: as already stated, a 'catch-all' type of survey that includes almost everything of conceivable importance will usually result in data of substandard quality and indeterminate analysis. Additional problems arise because some of the benefits of a project may be of an indirect kind. Attitudes may change; and 'psychic' income and costs may be relevant.

Jahnke, writing about the monitoring of livestock production systems, states that 'the objective should be to develop simple, quick, inexpensive methods of collecting data which are specific, objective, replicable and amenable to ... elementary analytical procedures'.[3] There is a growing awareness that certain variables are almost impossible to measure accurately, without incurring great cost: that 'Data collection should be limited to items that can be measured reasonably well within financial and human resources that are available.'[4] The UN carry the matter further stating that:

Indications may be direct, such as those cited above (usually monitoring indicators), or indirect (proxy). Indirect indicators (usually impact indicators) are used where direct measurement is not feasible or cost effective.[5]

The measures required are, of course, inextricably linked with the type of project under examination. They have to be easily defined and collected. Some should be closely related to project objectives, and others should provide general indicators of change. A trade-off between different criteria will often be required as the following illustration suggests. Data on income will not be easy to collect. A change in expenditure may not be a project objective, but it may be easier to define and measure, and can serve as a proxy for the change in income, which is a project objective.

The nutritional status of children is often not directly related to project objectives but it can serve as a general indicator of change in welfare levels; and the necessary data are easily collected. We discussed general problems of cause–effect assessment in the previous section. What framework and what indicators can we propose to isolate in practice the effect that can be attributed to the project from other effects that are due to exogenous factors?

A development project operates in an area, and with a community, subject to a wide variety of influences, climatic, economic, social, and governmental: everything is linked to everything else. In agricultural projects productivity levels are subject to a wide range of influences, many of which are outside the control of management. Rainfall will be a major factor in non-irrigated areas; and relative changes in crop prices decided upon at a national level may be more influential in farmers' decisions than the project interventions. The provision of a health clinic may lead to the population's seeking and obtaining qualified medical assistance when ill—the use of the clinic can be measured easily enough, and this shows the need for it. But suppose that the overall health status of the population shows little change: do we attempt to measure the comfort the population obtains from knowing that help is nearer at hand than hitherto? Or suppose that there is a marked improvement in child health, but that, in addition to the new clinic, there has been an active national campaign using the media and additional social workers to educate mothers about child care. How much of the local improvement is due to the clinic? Again, the construction of a rural market will rarely occur in isolation from other agricultural development efforts. An improvement in the production and marketing of food crops may be due, at least partly, to the availability of the market, but unusually favourable climatic conditions, an improved supply of subsidized inputs, and government announced changes in guaranteed producer prices may have played a greater part. How then do we evaluate the market?

There are two steps that need to be undertaken in a rigorous evaluation: namely the measurement of change and the attribution of a part or whole of this change to the intervention of the project. Merely estimating change can be difficult in a system that is subject to large variations from year to year or season to season. It may take many years to measure the trend in rural incomes due to fluctuations in climate—the occurrence of drought years and years of plenty. Attribution of causality of change is even more difficult to achieve. Indeed, in many cases we consider that the ambitions set for such evaluations are excessive. Surveys have been commissioned on a large scale and with considerable resources that nevertheless failed to achieve even reliable measures of change, whether project induced or not. Rigorous evaluation can be done, but should be attempted on a selective basis, not as a routine in each and every project. In most cases plausible inferences must be drawn from the time series of such indicators as adoption rates, especially repeat adoption rates, that indicate a measure of customer satisfaction. Further review of such monitoring data can be supplemented by case studies of a few project participants to establish through in-depth interview and observation the impact of the interventions on a case-by-case basis. At the very least, as we remarked in Chapter 5, such case studies may allow the evaluator to disprove an existing generalization.

For a full evaluation we have outlined how experimental methods suggest the use of a 'control' group of respondents, similar to the population in the impact area, subject to the same exogenous influences, and only differing in that the specific development stimulus under evaluation is missing; and we have indicated that identification of such a control group may prove difficult. The very fact that an adjacent area was omitted from the development project, rendering it as apparently suitable for the selection of 'control' respondents, may indicate that the facilities on offer would not have been cost-effective in such an area. The experience of those who have adopted this approach is daunting. Two quotations illustrate the point:

control groups have two important disadvantages. The first is that it is very costly and difficult to find farms (or households etc.) which match on a wide variety of characteristics but differ in just the one which is to be studied. ... The second problem is a conceptual one dealing with the choice of which factors are to be controlled and which factors are to be left uncontrolled.[6]

and

PRODESCH, a socio-economic programme in the Chiapas highlands of Mexico had by 1973 worked on about 300 of the 600 villages. ... In theory, it would be possible to estimate the impact ... by comparing the 300 worked villages ... with the 300 non-worked villages. However, the 300 worked villages tended to be close to major population centres, larger and more involved in the state's economy, while the 300 non-worked villages tended to be relatively inaccessible, smaller and less involved in the monetary economy.[7]

The use of a control group also implies the allocation of a considerable amount of the survey resources to this group. Often the control sample chosen is much smaller than that for the project area, but this may make little statistical sense. An attempt is to be made to detect a significant difference between changes in the before and after means of two groups; high sampling errors in any of the four surveys will be equally detrimental to this objective. About half the survey resources may therefore be required for the measurement of the control, and that group may later turn out not to have been a proper control at all. Even if a satisfactory control group is obtained the confounding effect of the other variables may well remain serious. The situation is analogous to testing the difference between two 'treatments' in a situation where the random effect, or residual variance, is much larger than the 'treatment' effect. An evaluation of an agriculture extension project in India[8] devoted great care to the choice of two areas (one subject to the new extension service, the other not) divided by a major river with little communication between the farmers on either side due to the absence of a bridge in the precise location chosen. The cropping patterns and farming systems were very similar and the microclimates identical. Even so the evaluation required the use of very advanced econometric analyses in order to factor out the possible effect of the new extension method. This example shows that the use of a control group is not always impracticable, but it will never be easy to design or implement such a survey.

If the country is carrying out a regular survey programme, it may be possible to utilize this as, at least, a partial substitute for specially selected control groups. If the project areas do not already appear in the national sample, additional selections can be made. In addition to the main indicators in the national sample, measured in the same way, a small number of supplementary specific indicators can be included in the project area and selected similar areas, but they should not be so numerous as to alter the character of the surveys.

This has the advantage of providing a time series, which assesses change more effectively than a single before–after comparison. Comparisons can be made between changes in the project area and in (a) similar places in the sample nearby, (b) in the region, and (c) in the nation. These comparisons should not be limited to average figures; where samples are retained over time the frequency distributions of changes for individual farmers should also be set side by side.

There is always a danger that concentration on the project and its direct effects may obscure some deleterious results. The control of water in an irrigation scheme may remove customary watering holes for nomadic groups; the extension of fish farming may spread bilharzia. The possibility of bad as well as good effects must be kept in view; and, since they may occur in unexpected ways, special effort is required to see that they do not escape notice. 'Project blindness' must be avoided, and this is one reason for insisting that any evaluation team must include persons not involved with the project.

14.4 SURVEILLANCE AND INFORMATION SYSTEMS

Partly as a result of the 1974 World Food conference, considerable attention has been focused on surveillance and information systems. The FAO now operates a Global Food Information System and several international agencies are promoting nutritional surveillance surveys. To some extent, this is just a renaming of current and longitudinal studies which may be combined with forecasting, so making possible the maintenance of up-to-date situation reports regarding the national position. The danger in some of the recent promotional efforts is that the recommended systems may be seen as something distinct and separate from normal statistical activities. This attitude leads inevitably to a duplication of effort, a proliferation of data collection agencies, and consequent waste of scarce statistical resources. No development of separate information systems is normally required since surveillance can usually be integrated into existing effective survey programmes.

The level of food stocks, the likely size of the harvest for the current season, and food consumption levels of the population are basic statistics needed in any country. The attempt to provide them on a continuous basis may be considered as the fundamental surveillance and information system. Reports of food stocks by main stock holders have to be supplemented by estimates of on-farm stocks. These are difficult to estimate unless there is a continuous household, and hence holding, survey programme. The existence of such a programme is a necessary, and usually also a sufficient, condition for the running of a surveillance system.

The early estimation of harvest prospects, which can be provided for in such a programme, is covered in the following section. In the early years, a combination of simple meteorological data, direct observations from crop cuts or counts, and interviews with a national sample of farmers should meet the forecasting needs of a surveillance system.

The more difficult data to collect are those relating to food consumption; the difficulties are outlined in Chapter 11. Food consumption surveys will not usually be part of a constant surveillance system. This is not to say that such surveys do not have a place in a cycle of surveys over several years, but they need not be an annual feature. It will usually be easier to measure directly the nutritional status of the population. A person's nutritional status reflects his past history of consumption as well as his current food intake and any deterioration or improvement in the food supply position is likely to be reflected fairly quickly in measurements of the affected population. Nutritional status is admittedly affected by factors other than absolute levels of food consumption, and obviously interacts with the general health of the individual. But this is all to the good as far as the surveillance system is concerned; although not a specific indicator of the level of food consumption, nutritional status may be said, in a layman's generalization, to indicate the level of food utilization. Abundant milk supply is of little use if it causes upset stomachs in the children who drink it.

A full measurement of nutritional status requires a combination of clinical, biochemical, and anthropometric observations. Moreover, the current position, or recent changes in the food consumption and health situation, are more readily reflected in data on children than on adults. The nutritional status of an adult reflects a long history of feeding, and adults may adapt to temporary changes in their situation with little apparent effect on their bodies. Clinical and biochemical studies require scarce and expensive expertise. This situation provides excellent occasion for co-operation between official statistics agencies and specialist surveyors either in or out of government. The national sample survey can include a simple nutritional status study, limited to young children; the specialist surveyor can supplement these data by case studies involving more rigorous methodology and covering a wider range of ages.

Simple indicators of nutritional status are height and weight for a given age. The necessary measurements are relatively easy to take. The distributions of height for age, weight for age, and weight for height of a young population enable nutritionists to deduce a good deal, especially when supplemented by other data. The length of breast-feeding, the main ingredients of the solid diet, recent or current illnesses (albeit in lay terms) may be obtained by interviewing the mother of the child. Moreover, a nutritional status survey forming part of a household survey programme links the data on the child with data showing the socio-economic conditions of the household of which it is part. Extra questions on a household or a community basis can be added dealing with such matters as sanitation, water, and other services. Figure 14.1 provides an example of a nutritional status questionnaire included in a national household survey programme. A module such as this, repeated at regular intervals, provides the basis of a nutritional surveillance system. If the results are put together with data about the storage and disposal of food crops, a food information and surveillance system is provided at little extra cost.[9]

FORM NUT/1.

CENTRAL BUREAU OF STATISTICS: KENYA GOVERNMENT

CONFIDENTIAL

PROVINCE_____

LOCATION/TOWN_____

RURAL/URBAN SURVEY 1978-79

NUTRITION MODULE

TARGET POPULATION: All Children in sample households
between 6 months and 5 years (i.e. 60 months) of age

DATE OF THIS VISIT............

CODE | 1 | 2 | 3 | 4 | 5 | 6 | 7 | 8 |

A. CHILD MEASUREMENTS

Columns: 1 SERIAL NUMBER | 2 SEX (1=M 2=F) | 3 DATE OF BIRTH (DAY MONTH YEAR) | 4 AGE In months | 5 BIRTH ORDER | 6 MONTHS WHEN BREASTFEEDING | 7 AGE WHEN OF SUPPL. FEED-ADDED In | 8 WEIGHT In Kilos | 9 HEIGHT Length in Cms | 10 FOOD: OUT OF WHAT IS PORRIDGE MADE? / WHAT IS ADDED TO PORRIDGE? / EVER USED COMMERCIAL BABY FOOD? | 11 SICKNESS: WAS CHILD SICK IN LAST 2 WEEKS? / TYPE OF SICKNESS? / ACTION TAKEN / WAS FOOD WITHDRAWN?

CODES

Out of what is porridge made?
0=Maize only
1=Millet only
2=Maize & Millet
3=Cassava only
4=Cassava, Maize & Millet
6=Bananas
7=Potatoes
8=Other...........

What is added to porridge?
0=Nothing
1=Milk
2=Sugar
3=Milk & Sugar

Ever used Commercial baby food?
1=Yes
2=No

Was child sick in last 2 weeks?
0=No
1=For 1 day
2=For 2-3 days
3=For 4-7 days
4=For over 1 week

Type of Sickness
1=Fever
2=Diarrhea
3=Fever & Diarrhea
4=Other

Action Taken
1=Taken to health centre or disp.
2=Taken to hosp. or private doctor
3=Purchased tablets
4=Used traditional medicine
5=No treatment

Was food withdrawn?
1=Yes
2=No

B. HOUSEHOLD SOCIAL AMENETIES

WATER SUPPLY

WET SEASON
SOURCE:
1=Still Pond
2=Small dam
3=Stream
4=Spring
5=Well
6=Borehole
7=Sub-surface dam
8=Jabias rain water
9=Piped water
0=Other

SINCE WHEN IS SOURCE AVAILABLE? In Months Kms

Columns 11 12 13 14 15

DRY SEASON
SOURCE:
1=Still Pond
2=Small dam
3=Stream
4=Spring
5=Well
6=Borehole
7=Sub-surface dam
8=Jabias rain water
9=Piped water
0=Other

SINCE DIST-WHEN ANCE IS SOURCE AVAIL-ABLE? In Months Kms

Columns 16 17 18 19 20

SEWAGE DISPOSAL
0=None
1=Main Sewer
2=Septic Tank
3=Pit Latrine
4=Bucket Latrine
5=Cess Pool

Column 21

RADIO
DO YOU LISTEN TO THE RADIO?
1=Yes
2=No

Column 22

IF YES: WHAT IS THE SOURCE?
1=Own
2=Social Centre
3=Neighbour
4=Other

Column 23

CARD N | 2 | | |

FIG. 14.1 *Source:* as in Fig. 6.1

14.5 FORECASTING SURVEYS

Planners require forecasts for many sectors of the economy: businessmen base their budgets on forecasts of sales, and researchers study communities in order not only to describe their condition but to forecast their future behaviour if offered certain stimuli. Many forecasts are based on extrapolation of existing data series, or on the basis of simulation runs, using a model that adopts certain assumed or empirically derived relationships between selected variables. We are primarily concerned in this book with forecasts made from survey data collected for this specific purpose.

The initials KAP have been used to describe surveys that investigate respondents' knowledge of, attitude to, and practice of contraceptive use. This acronym could be applied to many surveys intended to result in forecasts. The survey should be designed to probe the respondent's knowledge of the current situation, his attitude to the factors influencing this situation, and his usual practice to aid or counteract these factors, depending on whether he regards them as beneficial or not. From these facts a forecast of the eventual result may be made. Interviews may be linked with objective measurements of certain aspects of the current situation, and may close with a request to the respondent to provide his own forecast.

The scientist reared on controlled experimentation and a wide range of direct observations might regard these procedures as too subjective to be of much use. However, in studying agriculture, business, or social behaviour, it is a mistake to underestimate the knowledge of those so occupied. Their evaluation of the impact of the factors at work may be a good deal sounder than that deduced from models relying entirely on apparently objective observations. A network of observation posts and recording instruments and an army of measurers may result in forecasts little better than those achieved by interviewing the population involved. We are not of course suggesting that farmer interviews substitute for scientific meteorological forecasting; but a developing country may well meet its crop forecasting needs in a cost-effective way by combining interviews with simple meteorological observations, rather than by adopting a system requiring a mass of accurately measured data, which may well not be analysed in time.

The forecasting of crop production is, indeed, one of the commonest requirements. An approach using satellite imagery, aerial photography, and extensive ground observation networks may be appropriate for countries such as the United States of America, but it is inappropriate for most developing countries, and not only for reasons of cost. Remote sensing at its present stage is not adapted to small-scale cultivation and mixed cropping.

Production of crops that are particularly important to the country's economy can be forecast, more or less successfully, by maintaining a national sample of 'blocks' containing the crop and taking vital measurements at appropriate times before the harvest. For example, the size of the cocoa crop in Ghana was forecast for some years by measuring, on a sample of blocks of cocoa trees, the circumference and diameter of the cocoa pod at a certain stage of its growth.

In the absence of such objective surveys a great deal, nevertheless, can be achieved. A sample of farmers may already exist, or can be set up and used to obtain a crop forecast by means of an interview during the growing season. The essence of such a crop forecast is that it should be made quickly. To collect the data one month before the harvest, but to take several months to process the data, defeats the purpose of the exercise. The normal procedures used by the surveyor for collection, transmission, editing, processing, and analysis of the data must be speeded up for the crop forecast. Close involvement of senior statistical staff is required in order that prompt and appropriate decisions can be made. The following steps need to be taken:

(a) the questionnaire mus be simple;
(b) non-response should be measured and reported, but must not be allowed to cause delay;
(c) completed questionnaires must be mailed direct from enumerator to surveyor;
(d) editing and validation must be in the nature of a single quick scan, with 'rule-of-thumb' adjustments;
(e) processing arrangements, including necessary computer programming, must be completed in advance;
(f) the sample weighting may have to be approximate;
(g) analyses should be restricted to simple distributions and aggregates;
(h) data release should be immediate to an approved circulation list.

The brief crop forecast questionnaire should consist of a limited number of questions capable of being readily answered, if only by an approximation to the unknown truth; opinions relating to the future must be expressed in either numeric or 'like this but not like that' terms. A crop forecast interview is not the time for in-depth questioning and probing of response accuracy. With speed as the overriding consideration, both in data collection and analysis, simplicity and brevity should be the keynotes. Questions that may be answered without difficulty include: Did you plant ... this season? Date of planting? Probable date of harvest?

The accuracy of dating sought depends upon the use to which the information is to be put and the extent that farmers will be able to give exact dates. Some farmers may be able to give the week of planting, others may refer to particular features of the rainy season. Questions relating to the extent of the planting are less easy for the respondent to answer when the crop is grown in small plots of unknown size. Trained enumerators

CENTRAL BUREAU OF STATISTICS: KENYA GOVERNMENT

CROP FORECAST - DECEMBER 1979

AREAS WITH TWO SEASON PLANTATION

PROVINCE________ DISTRICT ________

CLUSTER________ ENUMERATOR________

SUPERVISOR________ DATE________

CODE [1 2 3 4 5 6 7 8 9 10 11 12 13] HOLDING AREA-ACRES [14 15 16 17]

CROP	FIRST RAIN CROP HARVEST - CROP PLANTED IN THE FIRST HALF 1979								SECOND RAIN CROP - CROP PLANTED IN THE SECOND HALF 1979						
	DID YOU PLANT CROP THIS YEAR? 1=YES 2=NO	IF YES							DID YOU PLANT CROP THIS SEASON? 1=YES 2=NO	IF YES					
		SIZE OF HARVEST		MONTH OF HARVEST	AMOUNT SOLD		TO WHOM 1=MAIZE & P.BOARD 2=LOCAL TRADER 3=OTHERS	UNIT SELLING PRICE		DID YOU PLANT 1=MORE AREA 2=SAME AREA 3=LESS AREA THAN LAST SEASON?	SIZE OF ACTUAL AND EXPECTED HARVEST		SIZE OF ACTUAL AND INTENDED SALE		EXPECTED MONTH OF HARVEST
		BAGS	DEBES		BAGS	DEBES		SHS			BAGS	DEBES	BAGS	DEBES	
MAIZE 0 1 (18 19)	20	21 22	23 24	25 26	27 28	29 30	31	32 33 34	35	36	37 38	39 40	41 42	43 44	45 46
BEANS 2 3 (18 19)	20	21 22	23 24	25 26	27 28	29 30	31	32 33 34	35	36	37 38	39 40	41 42	43 44	45 46
ENGLISH POTATOES 3 0 (18 19)	20	21 22	23 24	25 26	27 28	29 30	31	32 33 34	35	36	37 38	39 40	41 42	43 44	45 46
TOMATOES 3 9 (18 19)	20	21 22	23 24	25 26	27 28	29 30	31	32 33 34	35	36	37 38	39 40	41 42	43 44	45 46
MILLET 0 7 (18 19)	20	21 22	23 24	25 26	27 28	29 30	31	32 33 34	35	36	37 38	39 40	41 42	43 44	45 46
SORGHUM 1 1 (18 19)	20	21 22	23 24	25 26	27 28	29 30	31	32 33 34	35	36	37 38	39 40	41 42	43 44	45 46

FIG. 14.2 *Source*: as in Fig. 6.1

who have been working in the area may be able to give approximations of area within certain size classes.

The farmer's estimate of the likely size of the harvest will often be in terms of a standard unit, such as a bag, or a standard tin commonly used for marketing purposes. If necessary, the estimate of size can be approached less directly by asking him to estimate this year's potential relative to the size of last season's harvest, calculating the percentage increase or decrease in the office. The interview may also include questions on the respondent's intention to sell part or all of the harvest and the type of buyer to whom he intends to sell.

The need for speed does not allow lengthy recall procedures to be incorporated for those 'missed' from the survey in the initial sweep across the area. A higher rate of non-response than that aimed at in a current statistics survey must be tolerated. For this reason, sample weights may have to be computed quickly allowing for this non-response, despite its implications in terms of the accuracy of the estimate.

Rapid processing of the data is needed, with little emphasis on detailed and comprehensive validation, followed by the preparation of simple analytical tables. The first requirement is a forecast that at least indicates whether or not there is a positive or negative change since the last season, and if so whether it is likely to be a major one. Subsequently, the data can be polished and a more refined forecast issued.

An example of an interview-based, crop forecast questionnaire is shown as Figure 14.2. This survey resulted in early forecasts of the production of the three crops that proved later to be consistent with a slower survey based on plant counts and cob size (maize only) and with subsequent estimates of actual production. It is doubtful if greater accuracy would have been achieved by measuring various agrometeorological variables, and the latter approach would certainly have resulted in much greater expenditure and probably a later forecast. However, indirect forecasts using a simple regression of yield on, say, rainfall, planting date, and growing period may provide useful supplementary evidence.

Other forecasting surveys, for example, business expectation and expected manpower requirements, may be spread over a longer time period for data collection and analysis, so that greater deliberation can be exercised at each stage of the survey. The wording of the questions must, of course, still be chosen with care, and should be directed to actions taken and not to vague expressions of intent.

Notes

1. Casley and Kumar, *Project Monitoring and Evaluation in Agriculture*.
2. Moser, C. A., and Kalton, G., *Survey Methods in Social Investigation*, Heinemann, London, 1971. Chapter 9 provides a useful discussion of the issues touched on in this section.

3. Jahnke, H. E., 'Monitoring Livestock Development Projects: The Approach of the International Livestock Centre for Africa', prepared for World Bank Regional Workshop, Nairobi, 23–27 April 1979.
4. Cernea, M., and Tipping, B., 'A System for Monitoring and Evaluating Agricultural Extension Projects', *World Bank Staff Working Paper*, Washington, DC, Dec. 1977.
5. UN ACC Task Force on Rural Development, *Guiding Principles for the Design and Use of Monitoring and Evaluation in Rural Development Projects and Programmes*, Rome, 1984, p. 37.
6. Daines, S. R., 'An Overview of Economic and Data Analysis Techniques for Project Design and Evaluation', USAID, Washington, DC, Aug. 1977.
7. *Systematic Monitoring and Evaluation of Integrated Development Programmes, A Source Book*, UN, New York, 1978.
8. Feder, G., Lau, L., and Slade, R., *The Impact of Agricultural Extension: A Case Study of the Training and Visit System in Haryana, India*, World Bank, 1985.
9. Third Rural Child Nutrition Survey 1982, Republic of Kenya, Central Bureau of Statistics, Nairobi, 1983.

Index